CLINICAL CHEMISTRY

MULTIPLE CHOICE QUESTIONS

William J Marshall MA MSc PhD MB BS MRCP(UK) FRCPath

Senior Lecturer in Clinical Biochemistry
King's College School of Medicine and Dentistry
University of London
London, UK

Hon. Consultant in Clinical Biochemistry
King's College Hospital
London, UK

D1384004

Gower Medical Publishing • London • New York

Distributed in USA and Canada by:
JB Lippincott Company
East Washington Square
Philadelphia, PA 19105
USA

Distributed in Southeast Asia, Hong Kong, India and Taiwan by:
APAC Publishers Services
30 Jalan Bahasa
Singapore 1129

Distributed in the UK and Continental Europe by:
Gower Medical Publishing
Middlesex House
34–42 Cleveland Street
London, W1P 5FB
UK

Distributed in Japan by:
Nankodo Co Ltd
42-6 Hongo 3-chrome
Bunkyo-ku
Tokyo 113
Japan

Distributed in Australia and New Zealand by:
Harper Educational (Australia) Pty Ltd
PO BOX 226
Artarmon NSW 2064
Australia

Distributed in South America by:
Harper Collins Publishers Latin America
701 Bricknell Avenue
Suite 1750
Miami, Florida 33131
USA

Library of Congress Cataloging in Publication Data and
British Library Cataloguing in Publication Data are available

ISBN 0-379-44729-9

Publisher	Fiona Foley
Project Editor	David Cooke
Design	Timothy Read
Production	Susan Bishop
	Adam Phillips

Typeset on Apple Macintosh ®
Film output by The Text Unit, London, UK.
Text set in Helvetica and Times
Printed by The Bath Press, Bath, UK.

Contents

Preface

Although their merits continue to be debated, multiple choice questions have achieved an established place in examinations, providing as they do an easily assessed test of objective knowledge. They can also be used to guide students' revision, since reference to the answers will immediately indicate whether a question on a particular topic can be answered with confidence or whether further reading is required.

The questions in this book are based on the text of the author's *Illustrated Textbook of Clinical Chemistry, Second Edition.* The answers for the majority of questions can be found in the relevant chapters in the text. The last forty questions are based on data from clinical cases. Their successful completion often requires the amalgamation of information from several chapters.

Once attempted, a multiple choice question loses its novelty to the reader. Before attempting a particular group of questions, therefore, readers are advised to revise the subject of the questions. Their performance will then provide a more objective assessment of knowledge than if questions are attempted piecemeal.

The language used in multiple choice questions continues to cause confusion, and has been a source of much criticism. Every effort has been made to avoid ambiguity in these questions, but the author apologizes, and accepts full responsibility, for any that remains.

The questions are of the multiple determinate type - that is, any number of responses may be correct. In accordance with practice in examinations with which the author has been concerned, however, in no case are all the responses incorrect.

I am indebted to students at my own Medical School and on Professor David Slome's Primary FRCS course who have attempted and commented on many of these questions. In addition, my colleagues Dr Wassif Wassif, Miss Joan Butler and Dr Michael Norman have between them read every question. They have made many invaluable suggestions but any errors which remain are the responsibility of the author alone.

At Gower, I am grateful to Fiona Foley, who suggested that I write this book. I could not have wished for a better editor than David Cooke, and I thank Timothy Read for making a book of MCQs reader-friendly. At home, I thank Wendy for her encouragement and patience throughout its writing.

W.J.M.

1

Biochemical Tests in Clinical Medicine

1.1 The following statements about biochemical tests are correct:

a. They are only diagnostically useful in diseases of obvious endocrine or metabolic origin

b. They can be used to screen for complications of treatment

c. They can suggest the presence of a disease in the absence of any clinical evidence

d. They may provide information that helps in the management of patients even if the precise diagnosis remains uncertain

e. Two results obtained on different occasions in an individual will not indicate a clinically significant change if they both lie within the reference range for the population.

1.2 A precise analytical method is one which gives results which

a. Are comparable with results obtained for the same analyte using other methods

b. Are reproducible

c. Are simple to interpret

d. Are unaffected by the presence of other, similar substances in the sample

e. Discriminate reliably between health and disease.

1.3 An accurate analytical method is one which
   a. Can be relied on to detect small changes in the concentration of an analyte
   b. Gives results which are very close to the true values
   c. If used to measure an analyte in a sample of a given population, should produce a very similar mean result to the mean result for the whole of the population
   d. Is capable of measuring very small quantities of the analyte
   e. Is simple to perform.

1.4 If the distribution of test results in a population is Gaussian
   a. Approximately 95% of the population will have results within the range between the mean minus two standard deviations and the mean plus two standard deviations
   b. No results will be higher than the mean plus three standard deviations
   c. It can be assumed that the analytical method used is both precise and accurate
   d. It would be expected that approximately half the population will have results that are above the mean value
   e. The distribution of values for the population as a whole will be symmetrical about the mean.

1.5 The reference range for an analyte
   a. Excludes values which can occur in disease
   b. In a specific group of individuals is likely to be smaller than for the population as whole
   c. Includes all values likely to be found in healthy people
   d. Is an indication of the extent of variation in the concentration of an analyte likely to occur in a healthy individual
   e. May vary according to the method used for its measurement.

1.6 If a test result is inconsistent with the clinical findings
   a. It can be assumed that a mistake has occurred in the laboratory
   b. It is advisable to repeat the test on a fresh specimen if there is no obvious exlanation
   c. It suggests that the analytical method is inaccurate
   d. The clinical diagnosis may need to be reconsidered
   e. The result should be ignored.

1.7 The following statements about the sensitivity of a diagnostic test are true:
   a. It is a measure of the number of patients with a disease who are detected by the test
   b. It is decreased by false negative results
   c. It is higher if the disease has a high prevalence
   d. It is independent of the number of false positive results
   e. It should be maximized if the test is to be used to screen for a harmful condition.

1.8 A diagnostic test has high specificity if
   a. Most individuals who do not have the condition in question have a negative test result
   b. The analytical method is very precise
   c. The condition it is used to diagnose is rare
   d. The false negative rate is low
   e. The test gives abnormal results in only one condition.

1.9 An apparently healthy individual has a test result which is greater than the upper limit of the reference range for a particular analyte. It follows that
   a. Further investigations will be required to establish whether disease is present or not
   b. Sub-clinical disease is present
   c. The analytical method is imprecise
   d. The reference range is incorrect
   e. The test has low specificity.

1.10 The following characteristics are desirable in a test to be used for screening for a potentially serious, but presently asymptomatic, condition:
   a. Clear separation of values characteristic of presence and absence of condition
   b. High predictive value for a negative result
   c. High true positive rate
   d. Low analytical variation
   e. Low false negative rate.

2

Water and Sodium

2.1 The following contribute more than 10% to the osmolality of the plasma:
   a. Bicarbonate
   b. Glucose
   c. Potassium
   d. Protein
   e. Sodium.

2.2 The following statements concerning sodium are correct:
   a. Most of that which is secreted into the gut is usually excreted into the stool
   b. It is primarily an extracelluar ion
   c. Its distribution between the extra- and intracellular compartments is energy-dependent
   d. Its plasma concentration is an accurate guide to total body sodium content
   e. Renal reabsorption is stimulated by aldosterone throughout the nephron.

2.3 The following statements about body water in adults are correct:
   a. Approximately two-thirds is in the intracellular compartment
   b. It contributes approximately 60% of total body weight
   c. It contributes a greater percentage of the body weight than in infants
   d. Obligatory losses are approximately 1500ml/day
   e. The plasma contains approximately 8% of the total.

5

2.4 Physiological responses to loss of water from the extracellular fluid include
a. Early decrease in glomerular filtration rate
b. Inhibition of vasopressin secretion
c. Movement of water from the intracellular space
d. Natriuresis
e. Thirst.

2.5 The following can stimulate vasopressin secretion:
a. Angiotensin II
b. Carbamazepine
c. Decreased osmolality of the extracellular fluid
d. Hypervolaemia
e. Stress.

2.6 Causes of predominant water depletion include
a. Adrenal failure
b. Burns
c. Hyperventilation
d. Small intestinal fistula
e. Treatment with thiazide diuretics.

2.7 In a patient with fluid loss, the following would suggest that the loss was approximately isotonic with respect to plasma:
a. Intense thirst
b. Plasma urea concentration within the reference range
c. Postural hypotension
d. Production of a highly concentrated urine
e. Tachycardia.

2.8 Early features of pure water depletion include
a. Decrease in skin turgor
b. Decrease in urine volume
c. Increased aldosterone secretion
d. Increased haematocrit
e. Increase in plasma urea concentration.

2.9 Pathological sodium retention is a recognized consequence of
a. Cortisol deficiency
b. Nephrotic syndrome
c. Renal artery stenosis
d. Treatment with chlorpropamide
e. Conn's syndrome.

2.10 Recognized causes of hyponatraemia include
a. Anterior pituitary failure
b. Acute renal failure
c. Rapid intravenous infusion of an isotonic dextrose solution
d. Untreated diabetes insipidus
e. Hypothyroidism.

2.11 The following are characteristic of the syndrome of inappropriate antidiuresis:
a. Increased plasma osmolality
b. Increased plasma urea concentration
c. Peripheral oedema
d. Urine osmolality inappropriately low in relation to plasma osmolality
e. Urine sodium concentration <20mmol/l.

2.12 Conditions associated with the syndrome of inappropriate antidiuresis include
a. Adrenal failure
b. Cystic fibrosis
c. Head injury
d. Intracranial tumours
e. Tuberculosis.

2.13 A patient has a plasma sodium concentration of 124mmol/ and osmolality of 260mmol/Kg. Plasma urea and glucose concentrations are normal. Possible underlying conditions/causes include
a. Ectopic secretion of antidiuretic hormone
b. Intravenous infusion of mannitol
c. Severe hypertriglyceridaemia
d. Severe sodium depletion
e. Water depletion.

2.14 A clinically dehydrated patient has a plasma sodium concentration of 130mmol/l; the urine sodium concentration is 80mmol/l. Likely causes include
a. Adrenal failure
b. Pre-renal renal failure
c. Primary aldosteronism
d. Severe diarrhoea
e. Sodium-losing nephropathy.

2.15 Hyponatraemia with evidence of extracellular fluid overload is a recognized finding in patients with
a. Acute renal failure
b. Chronic liver disease
c. Congestive cardiac failure
d. Cushing's syndrome
e. Nephrotic syndrome.

2.16 Investigations of potential value in the investigation of a patient with hyponatraemia include
a. Measurement of plasma creatinine concentration
b. Measurement of plasma glucose concentration
c. Measurement of sweat sodium concentration
d. Measurement of urine osmolality
e. Short Synacthen test.

2.17 Hypernatraemia is a recognized finding in patients
a. Treated with osmotic diuretics
b. Who are unable to drink
c. With fever
d. With hypopituitarism
e. With untreated diabetes insipidus.

3

Potassium

3.1 Potassium

    a. Can be conserved by the kidneys to a greater extent than sodium

    b. Concentration in plasma is higher than in serum

    c. Depletion is frequently associated with alkalosis

    d. Is a predominantly intracellular cation

    e. Transport into cells is stimulated by insulin.

3.2 Potassium excretion by the kidney is increased by

    a. Aldosterone

    b. Increased distal tubular sodium reabsorption

    c. Osmotic diuresis

    d. Treatment with spironolactone

    e. Vasopressin.

3.3 Recognized causes of hypokalaemia include

    a. Crush injuries

    b. Parenteral feeding

    c. Systemic acidosis

    d. Treatment with cytotoxic drugs

    e. Treatment of pernicious anaemia.

3.4 Recognized features of hypokalaemia include

    a. Broadening of QRS complex on electrocardiogram

    b. Constipation

    c. Depression

    d. Increased muscle tone

    e. Secondary aldosteronism.

3.5 Recognized causes of hyperkalaemia include
a. Acute renal failure
b. Conn's syndrome
c. Diabetic ketoacidosis
d. Mismatched blood transfusion
e. Treatment with thiazide diuretics.

3.6 The potassium concentration in a serum sample is found to be 6.1mmol/l. Renal function was normal. The following additional findings would suggest that this was not a true measure of the *in vivo* concentration:
a. Calcium concentration 0.8mmol/l
b. Glucose concentration 1.0mmol/l
c. Red discolouration of serum
d. Phosphate concentration 3.6mmol/l
e. Sodium concentration 150mmol/l.

3.7 Means of reducing a high plasma potassium concentration include giving
a. Calcium gluconate intravenously
b. Calcium resonium orally
c. Dexamethasone orally
d. Glucose and insulin intravenously
e. Sodium bicarbonate intravenously.

4

Hydrogen Ion Homoeostasis and Blood Gases

4.1 Processes generating hydrogen ions include
a. Anaerobic respiration
b. Gluconeogenesis from lactate
c. Glycolysis
d. Ketogenesis
e. Oxidation of sulphur-containing amino acids.

4.2 The following statements about hydrogen ions in the body are correct:
a. Intracellular concentration is normally lower than extracellular
b. Renal excretion is approximately 60mmol/24h
c. The amount of carbon dioxide produced by the body is equivalent to 15mmol hydrogen ion/24h
d. The normal blood concentration is approximately40nmol/l
e. They are buffered in the extracellular fluid mainly by phosphate.

4.3 The following statements about the handling of hydrogen ions and bicarbonate in the kidney are correct:
a. Filtered bicarbonate is reabsorbed mainly in the distal convoluted tubule
b. Hydrogen ions filtered by the glomerulus make a major contribution to the amount excreted in the urine
c. The minimum attainable urinary pH is 1.5
d. The process of bicarbonate reabsorption requires the generation of bicarbonate in renal tubular cells
e. The reabsorption of filtered bicarbonate results in net acid excretion.

4.4 Carbon dioxide
    a. Combines readily with water to form carbonic acid
    b. Has a normal concentration in solution in arterial blood of less than 2mmol/l
    c. Has a normal partial pressure in arterial blood of approximately 5.5kPa
    d. Is irreversibly converted to bicarbonate in red blood cells
    e. Is required for the reabsorption of bicarbonate by renal tubular cells.

4.5 Arterial hydrogen ion concentration
    a. Is directly proportional to the partial pressure of carbon dioxide ($P\text{co}_2$)
    b. Is increased by a decrease in $P\text{co}_2$
    c. Is inversely related to bicarbonate concentration
    d. Is on average 10nmol/l higher than in venous blood
    e. Should be measured in blood anticoagulated with citrate.

4.6 Recognized clinical features of an elevated arterial partial pressure of carbon dioxide ($P\text{co}_2$) include
    a. Cyanosis
    b. Drowsiness
    c. Muscle cramps
    d. Papilloedema
    e. Vasodilatation.

4.7 Causes of non-respiratory (metabolic) acidosis include
    a. Chronic carbon dioxide retention
    b. Ethylene glycol poisoning
    c. Potassium depletion
    d. Prolonged nasogastric aspiration
    e. Tissue hypoxia.

4.8 The following would be expected in a patient with a non-respiratory (metabolic) acidosis due to acid ingestion
  a. Acidic urine
  b. Decreased plasma ionized calcium concentration relative to total calcium
  c. Low arterial partial pressure of carbon dioxide (P_{CO_2})
  d. Low arterial partial pressure of oxygen (P_{O_2})
  e. Low plasma bicarbonate concentration.

4.9 Lactic acidosis is a recognized complication of
  a. Glycogen storage disease type I
  b. Hepatic failure
  c. Intravenous fructose infusion
  d. Severe shock
  e. Treatment of non insulin-dependent diabetes mellitus with sulphonylureas.

4.10 Causes of non-respiratory (metabolic) acidosis associated with a normal anion gap include
  a. Acute renal failure
  b. Alcoholic ketoacidosis
  c. Lactic acidosis
  d. Pancreatic fistula
  e. Renal tubular acidosis.

4.11 Recognized causes of respiratory acidosis include
  a. Anxiety state
  b. Barbiturate overdose
  c. Guillain–Barré syndrome
  d. Severe asthma
  e. Strenuous exercise.

4.12 In respiratory acidosis

a. A low plasma bicarbonate concentration would suggest that there was also a non-respiratory (metabolic) component to the acidosis

b. A markedly elevated bicarbonate concentration would suggest that the condition was acute

c. Compensation involves renal hydrogen ion excretion

d. Compensation may restore the arterial partial pressure of carbon dioxide (P_{CO_2}) to normal

e. Hydrogen ions are buffered by haemoglobin.

4.13 Non-respiratory (metabolic) alkalosis

a. Can usually be completely compensated by hypoventilation

b. Is accompanied by a high plasma bicarbonate concentration

c. Is a recognized consequence of fluid loss through an ileostomy

d. Is often associated with hypokalaemia

e. Usually requires urgent treatment by the administration of ammonium chloride.

4.14 In a patient with vomiting and pyloric stenosis

a. Any disturbance of acid–base homoeostasis is usually only mild

b. Hyperkalaemia is usual

c. Pre-renal uraemia is a recognized finding

d. The presence of a low arterial partial pressure of carbon dioxide (P_{CO_2}) suggests that the disorder is of very recent onset

e. The urine is often alkaline in severe cases.

4.15 Respiratory alkalosis
   a. Is characterized by a decrease in the arterial partial pressure of carbon dioxide (P_{CO_2})
   b. Can cause paraesthesiae
   c. Can complicate head injury
   d. Can be compensated by decreased renal excretion of hydrogen ion
   e. Is usually associated with a marked increase in the arterial partial presssure of oxygen (P_{O_2}).

4.16 A patient is admitted to hospital unconscious. Analysis of arterial blood shows: [H$^+$] = 40nmol/l, P_{CO_2} = 3.0kPa. These data are compatible with
   a. Barbiturate poisoning
   b. Combined respiratory alkalosis and non-respiratory (metabolic) acidosis
   c. Diabetic ketoacidosis
   d. Normal acid–base status
   e. Salicylate poisoning.

4.17 The following statements concerning the interpretation of acid–base data are correct:
   a. A raised P_{CO_2} is always present in patients with respiratory acidosis
   b. Compensatory changes may mimic primary disturbances
   c. Detecting the presence of compensatory changes requires knowledge of the standard bicarbonate concentration
   d. The derived bicarbonate concentration provides essential additional information
   e. The presence of a normal arterial hydrogen ion concentration implies normal acid–base status.

4.18 Oxygen delivery to tissues can be decreased by
   a. Anaemia
   b. Low cardiac output
   c. Poisons which intefere with oxidative phosphorylation in the tissue
   d. Pulmonary fibrosis
   e. Vasoconstriction.

4.19 Causes of a low arterial partial pressure of oxygen (Po_2) include
   a. Congenital heart disease with left-to-right shunt
   b. Decreased affinity of haemoglobin for oxygen
   c. Decreased proportion of oxygen in inspired air
   d. Imbalance of pulmonary ventilation and perfusion
   e. Low barometric pressure.

5

The Kidney

5.1 Major functions of the kidneys include the
a. 1-hydroxylation of 25-hydroxycholecalciferol
b. Regulation of extracellular hydrogen ion concentration
c. Synthesis of creatinine
d. Synthesis of angiotensin
e. Secretion of erythropoietin.

5.2 The glomerular filtration rate depends on the
a. Blood pressure in glomerular afferent arterioles
b. Creatinine clearance
c. Hydrostatic pressure in the lumen of the nephron
d. Nature of the glomerular basement membrane
e. Number of functioning nephrons.

5.3 The glomerular filtration rate exerts a major influence on the plasma concentration of
a. Albumin
b. β_2-microglobulin
c. Glucose
d. Sodium
e. IgM.

5.4 Clinically useful indicators of glomerular function include the measurement of
a. Clearance of chromium-labelled EDTA
b. Plasma creatine kinase activity
c. Plasma potassium concentration
d. Urine osmolality
e. Urine pH.

5.5 The creatinine clearance
- **a.** Can only be measured if an accurate 24 hour urine collection can be made
- **b.** Increases during pregnancy
- **c.** Is directly proportional to plasma creatinine concentration
- **d.** Overestimates the glomerular filtration rate
- **e.** Tends to remain constant with increasing age.

5.6 Plasma creatinine concentration
- **a.** Can increase temporarily following meat ingestion
- **b.** Is determined in part by muscle bulk
- **c.** Is independent of hepatic function
- **d.** Is a sensitive indicator of mildly impaired renal function
- **e.** Tends to increase with increasing age.

5.7 Plasma urea concentration
- **a.** Can increase as a result of bleeding into the gut
- **b.** Can increase as a result of dehydration
- **c.** Is a more reliable indicator of renal function than plasma creatinine concentration
- **d.** Tends to fall in patients on low protein diets
- **e.** Within the reference range indicates normal renal function.

5.8 In a patient with oliguria, the following would suggest a pre-renal, rather than an intrinsic renal, cause
- **a.** Glycosuria
- **b.** Urine sodium concentration <10mmol/l
- **c.** Plasma osmolality 285mmol/kg
- **d.** Plasma potassium 5.2mmol/l
- **e.** Urine:plasma urea concentration ratio >20:1.

5.9 Causes of acute renal failure include
a. Haemorrhage
b. Hypocalcaemia
c. Septicaemia
d. Systemic lupus erythematosus
e. Treatment with aminoglycoside antibiotics.

5.10 Recognized features of acute oliguric renal failure include
a. Hyperkalaemia
b. Hypoglycaemia
c. Hypophosphataemia
d. Non-respiratory acidosis
e. Renal osteodystrophy.

5.11 Following the onset of the diuretic phase of acute renal failure
a. Any acidosis resolves rapidly
b. Hypercalcaemia is a recognized occurrence
c. The urine is usually highly concentrated
d. There is a risk of sodium depletion
e. There is usually an immediate fall in plasma creatinine concentration.

5.12 Characteristic features of end-stage renal failure include
a. Hyperalbuminaemia
b. Hypernatraemia
c. Hypotension
d. Macrocytic anaemia
e. Nocturia.

5.13 In a patient with renal failure, the following findings suggest that it is acute, rather than chronic:
a. Elevated plasma alkaline phosphatase activity
b. History of recurrent urinary tract infections
c. Kidneys of normal size
d. Peripheral neuropathy
e. Vascular calcification.

5.14 Factors implicated in the pathogenesis of renal osteodystrophy include
a. Decreased 25-hydroxylation of vitamin D
b. Hyponatraemia
c. Hyperphosphataemia
d. Secondary hyperparathyroidism
e. Systemic acidosis.

5.15 The following conditions are recognized causes of proteinuria
a. Diabetic nephropathy
b. Membranous glomerulonephritis
c. Multiple myeloma
d. Strenuous exercise
e. Urinary tract infection.

5.16 Recognized causes of nephrotic syndrome include
a. Minimal change glomerulonephritis
b. Orthostatic proteinuria
c. Renal amyloidosis
d. Renal tubular acidosis
e. The Fanconi syndrome.

5.17 Recognized complications of the nephrotic syndrome include
a. Hypertriglyceridaemia
b. Hypoglycaemia
c. Hypothyroidism
d. Increased susceptibility to infection
e. Vascular thrombosis.

5.18 Characteristic features of the Fanconi syndrome include
a. Alkalosis
b. Aminoaciduria
c. Galactosaemia
d. Glycosuria
e. Hyperphosphataemia.

5.19 Renal calculi can be composed primarily of
a. Calcium oxalate
b. Bililrubin
c. Cholesterol
d. Cystine
e. Uric acid.

5.20 Conditions associated with the formation of renal calculi include
a. Diabetes insipidus
b. Malabsorption
c. Primary hyperparathyroidism
d. Urinary tract infection
e. Working in hot environments.

6

The Liver

6.1 Functions of the liver include the synthesis of
a. Albumin
b. Coagulation factors
c. Glycogen
d. Immunoglobulins
e. Very low density lipoproteins (VLDL).

6.2 Tests that can be used to assess the function of the liver include the
a. Bromsulphthalein excretion test
b. Galactose tolerance test
c. Measurement of plasma alanine transaminase activity
d. Measurement of plasma transferrin concentration
e. Measurement of prothrombin time.

6.3 Bilirubin
a. In the conjugated form is highly toxic to the nervous system
b. In the plasma is normally mainly unconjugated
c. Is responsible for the normal colour of the urine
d. Is synthesized in hepatic parenchymal cells
e. When unconjugated, is insoluble in water.

6.4 Jaundice
a. Associated with dark urine is due to unconjugated bilirubin
b. In association with pale stools suggests biliary obstruction
c. Is always detectable in patients with Gilbert's syndrome
d. Is a sensitive indicator of the presence of liver disease
e. Is unlikely to be detectable clinically if the plasma bilirubin concentration is <50μmol/l.

6.5 Causes of primarily unconjugated hyperbilirubinaemia include
a. Biliary atresia
b. Cirrhosis
c. Dubin-Johnson syndrome
d. Haemolysis
e. Hepatitis.

6.6 Characteristic findings in adults with jaundice due to haemolysis include
a. Reticulocytosis
b. Bilirubinuria
c. Increased plasma haptoglobin concentrations
d. Markedly elevated plasma aspartate transaminase (AST) activity
e. Plasma bilirubin concentration usually less than 100μmol/l.

6.7 Recognized causes of acute hepatitis include
a. Gallstones
b. Ineffective erythropoiesis
c. Infection with cytomegalovirus
d. Persistence of infection with hepatitis B virus
e. Poisoning with carbon tetrachloride.

6.8 The following are usually present in patients with acute hepatitis
a. Elevated plasma aspartate transaminase (AST) activity
b. Elevated plasma creatine kinase activity
c. Elevated plasma immunoglobulin concentrations
d. Low blood glucose concentration
e. Low plasma albumin concentration.

6.9 In acute hepatitis
a. An elevation of plasma γ-glutamyl transpeptidase is diagnostic of an alcoholic aetiology
b. Bilirubinuria can be present at an early stage
c. Complete recovery is usual when the condition is due to hepatitis A
d. Plasma alkaline phosphatase activity is usually normal throughout the illness
e. The development of jaundice usually precedes any change in plasma enzyme activity.

6.10 Typical features of chronic active hepatitis include
a. Elevated plasma alkaline phosphatase activity
b. Elevated plasma aspartate transaminase (AST) activity
c. Increased plasma immunoglobulin concentration
d. Low plasma albumin concentration
e. Normal plasma bilirubin concentration.

6.11 Recognized causes of hepatic cirrhosis include
a. α_1-antitrypsin deficiency
b. Chronic infection with hepatitis B
c. Chronic pancreatitis
d. Paracetamol poisoning
e. Primary (idiopathic) haemochromatosis.

6.12 Recognized complications of hepatic cirrhosis include
a. Ascites
b. Encephalopathy
c. Gastrointestinal haemorrhage
d. Hepatocellular carcinoma
e. Malabsorption.

6.13 In hepatic cirrhosis
a. Hypoalbuminaemia usually occurs early in the course of the disease
b. Plasma γ-glutamyl transpeptidase activity is usually elevated only if alcohol is the cause
c. Prolongation of the prothrombin time not corrected by parenteral administration of vitamin K suggests impaired hepatic protein synthesis
d. The plasma alkaline phosphatase activity is typically increased to a greater extent than that of aspartate transaminase
e. The urine becomes dark in colour when jaundice develops.

6.14 In a patient with clinical and biochemical evidence of cholestasis, the following findings would suggest a specific diagnosis:
a. High plasma alkaline phosphatase activity
b. High plasma bile salt concentration
c. High serum titre of anti-mitichondrial antibodies
d. Low plasma caeruloplasmin concentration
e. Transferrin 100% saturated with iron.

6.15 The following statements about inherited disorders of bilirubin metabolism are correct:
a. Crigler-Najjar syndrome causes unconjugated hyperbilirubinaemia
b. Gilbert's syndrome is a harmless condition
c. In Dubin-Johnson syndrome, jaundice is due to unconjugated hyperbilirubinaemia
d. The more severe form of Crigler–Najjar syndrome is inherited as an autosomal recessive condition
e. The plasma bilirubin concentration typically increases during starvation in Gilbert's syndrome.

6.16 Recognized features of fulminant hepatic failure include
a. Hyperglycaemia
b. Hypernatraemia
c. Lactic acidosis
d. Low plasma urea concentration
e. Respiratory alkalosis.

6.17 Gallstones
a. Are a recognized complication of chronic haemolytic anaemias
b. Are usually composed of a single substance
c. Can cause unconjugated hyperbilirubinaemia
d. Often contain ammonium salts
e. Rarely contain cholesterol.

6.18 Wilson's disease
a. Can cause a haemolytic anaemia
b. Can present as a severe acute hepatitis
c. Has an autosomal dominant mode of inheritance
d. Typically causes a high plasma copper concentration
e. Usually presents in the fourth or fifth decade of life.

6.19 The following statements relating to drugs and the liver are correct:
- **a.** Halothane hepatotoxicity frequently causes chronic hepatitis
- **b.** Isoniazid toxicity typically causes cholestasis
- **c.** Phase 1 metabolism is responsible for the toxicity of paracetamol
- **d.** Rifampicin interferes with bilirubin uptake by the liver
- **e.** The hepatotoxicity of methyldopa is dose-related.

7

The Gastrointestinal Tract

7.1 Gastrin
a. Is a steroid hormone
b. Is secreted physiologically mainly by the pancreas
c. Is unstable in plasma in vitro
d. Secretion is normally inhibited by secretin
e. Stimulates gastric acid secretion.

7.2 Hypergastrinaemia is a recognized finding in
a. Chronic renal failure
b. Patients treated with inhibitors of gastric acid secretion
c. Patients who have undergone vagotomy
d. The majority of patients with uncomplicated duodenal ulceration
e. Zollinger–Ellison syndrome.

7.3 Functions of the pancreas include the secretion of
a. Bicarbonate
b. Bile salts
c. Chymotrypsinogen
d. Lipase
e. Pepsinogen.

7.4 Tests of pancreatic exocrine function include
a. Glucose tolerance test
b. Lundh test
c. Measurement of faecal fat excretion
d. Measurement of plasma amylase activity
e. Pancreolauryl test.

7.5 Recognized aetiological factors in acute pancreatitis include

a. Alcohol
b. Gall stones
c. Hypertriglyceridaemia
d. Hypocalcaemia
e. Infection.

7.6 Recognized findings in severe acute pancreatitis include

a. Hypercalcaemia
b. Hypoalbuminaemia
c. Decreased plasma lipase activity
d. Methaemalbuminaemia
e. Pre-renal uraemia.

7.7 In patients with chronic pancreatitis

a. Calcification is frequently visible on plain abdominal X-ray
b. Diabetes mellitus is a recognized complication
c. Effective treatment can usually reverse the underlying damage
d. Malabsorption occurs early in the course of the condition
e. The plasma amylase activity is usually increased.

7.8 The following statements concerning the absorption of nutrients are correct:

a. Bile salts are absorbed in the proximal jejunum
b. Disaccharides are absorbed without prior digestion
c. Enzymes secreted by the pancreas are essential for the absorption of glucose
d. Fat is absorbed in the terminal ileum
e. Vitamin B_{12} is absorbed in the terminal ileum.

7.9 Faecal fat excretion

a. Is decreased in patients with bacterial colonization of the small intestine
b. Is rarely increased in patients with intestinal disease causing malabsorption
c. Of more that 18mmol/24hours is considered abnormal
d. Should be measured over a period of at least three days
e. Should be measured while the patient is on a low fat diet.

7.10 The ^{14}C-triolein breath test

a. Is potentially unreliable in patients with chronic respiratory insufficiency
b. Is potentially unreliable in patients with diabetes mellitus
c. Requires measurements to be made over a three-day period
d. Requires the injection of ^{14}C-labelled triglyceride
e. Shows decreased excretion of $^{14}CO_2$ in patients with malabsorption.

7.11 Recognized features of generalized malabsorption include

a. Hypergammaglobulinaemia
b. Hypoalbuminaemia
c. Hypoglycaemia
d. Increased plasma alkaline phosphatase activity
e. Macrocytic anaemia.

7.12 Causes of generalized malabsorption include

a. Coeliac disease
b. Cystic fibrosis
c. Hartnup disease
d. Surgical resection of the descending colon
e. Zollinger–Ellison syndrome.

7.13 Tests of potential value in determining the cause of malabsorption include
  a. Jejunal biopsy
  b. Measurement of D-xylose absorption
  c. Prothrombin time
  d. Secretin-cholecystokinin test
  e. ^{14}C-triolein breath test.

7.14 The following tests are correctly paired with causes of malabsorption in which abnormal results would be expected:
  a. Measurement of breath hydrogen excretion following ingestion of lactulose: lactase deficiency
  b. Measurement of faecal radioactivity following parenteral administration of ^{51}Cr-labelled albumin: chronic pancreatitis
  c. ^{14}C-*p*-aminobenzoic acid (PABA) breath test: coeliac disease
  d. Schilling test: Crohn's disease
  e. ^{14}C-xylose breath test: bacterial colonization of small intestine.

7.15 In a patient with malabsorption, the following would suggest that intestinal disease, rather than pancreatic, was responsible:
  a. Abnormal lactose tolerance test result
  b. History of excessive alcohol intake
  c. Increased breath hydrogen excretion
  d. Low plasma calcium concentration
  e. Microcytic anaemia.

7.16 Coeliac disease

  a. Causes characteristic hypertrophy of intestinal villi
  b. Is caused by chronic parasitic infection of the small intestine
  c. Is a recognized cause of growth retardation
  d. Presents only in children
  e. Usually responds well to treatment.

7.17 The following statements about the short bowel syndrome are correct:

a. Gallstones are a recognized complication

b. Long term parenteral nutrition is likely to be required if less than 60cm of small intestine is present

c. More severe disturbances occur if the mid, rather than the proximal, jejunum has been resected

d. Urinary calculi are a recognized complication

e. Zinc deficiency is more likely if there is persistent diarrhoea.

8

The Hypothalamus and Pituitary Gland

8.1 Growth hormone
a. Action is mediated in part by somatostatin
b. Is an insulin agonist
c. Is a single chain polypeptide
d. Secretion is inhibited by somatomedin
e. Secretion is pulsatile.

8.2 Prolactin
a. Is a pro-hormone (the inactive precursor of a hormone)
b. Is a steroid hormone
c. Secretion is stimulated by dopamine
d. Secretion is stimulated by TRH (thyrotrophin releasing hormone)
e. Stimulates ovulation.

8.3 Thyroid-stimulating hormone (thyrotrophin, TSH)
a. Deficiency is a frequent cause of thyroid failure
b. Has a beta sub-unit identical with that of the gonadotrophins
c. Is a glycoprotein
d. Is rarely secreted by pituitary tumours
e. Secretion is suppressed in primary hyperthyroidism.

8.4 Regarding the gonadotrophins (follicle-stimulating hormone, FSH, and luteinising hormone, LH):

a. FSH secretion is inhibited by inhibin

b. Increased plasma concentrations are characteristic of primary gonadal failure

c. Ovulation is triggered by an increase in LH

d. The major function of FSH in men is to stimulate testosterone secretion

e. The plasma concentration of LH rises before that of FSH as the climacteric approaches.

8.5 Adrenocorticotrophic hormone (ACTH)

a. Has a physiological role in aldosterone secretion

b. Has melanocyte-stimulating activity

c. Is similar in structure to growth hormone

d. Is the precursor of the endorphins

e. Secretion is inhibited by cortisol.

8.6 Recognized causes of hypopituitarism include

a. Craniopharyngioma

b. Diabetes insipidus

c. Head injury

d. Sarcoidosis

e. Tuberculous meningitis.

8.7 Recognized features of hypopituitarism include

a. Anorexia nervosa

b. Hypocalcaemia

c. Hypoglycaemia

d. Hypertension

e. Hyponatraemia.

8.8 The following statements about pituitary tumours are correct:
   a. Gonadotrophin-secreting tumours are a frequent cause of infertility
   b. They are a recognized cause of diabetes insipidus
   c. They occur only in adults
   d. Tumours secreting ACTH (adrencocorticotrophic hormone) are usually associated with visual field defects
   e. Tumours secreting prolactin occur more frequently than tumours secreting growth hormone.

8.9 Recognized features of acromegaly include
   a. Elevated plasma concentration of IGF-1 (insulin-like growth factor 1, somatomedin-C)
   b. Excessive sweating
   c. Fasting hypoglycaemia
   d. Goitre
   e. Hypophosphataemia.

8.10 The following tests are of established value in the diagnosis of acromegaly:
   a. Dexamethasone suppression test
   b. Glucagon stimulation test
   c. Glucose tolerance test
   d. GnRH (gonadotrophin releasing hormone) test
   e. Measurement of growth hormone following infusion of arginine.

8.11 Hyperprolactinaemia is recognized to occur in patients with
   a. Chronic renal failure
   b. Hypothyroidism
   c. Non prolactin-secreting pituitary tumours
   d. Normal pregnancy
   e. Sheehan's syndrome.

8.12 Excessive secretion of prolactin can cause
a. Diabetes insipidus
b. Galactorrhoea
c. Gigantism
d. Hypothyroidism
e. Oligomenorrhoea.

8.13 Causes of cranial diabetes insipidus include
a. Craniopharyngioma
b. Hypercalcaemia
c. Lithium toxicity
d. Meningitis
e. Sickle cell disease.

8.14 Recognized features of untreated diabetes insipidus include
a. Hypokalaemia
b. Polydipsia
c. Early morning urine osmolality >750mmol/kg
d. Excessive thirst
e. Low plasma osmolality.

8.15 The following tests may assist in the differential diagnosis of polyuria:
a. Measurement of plasma calcium concentration
b. Measurement of plasma creatinine concentration
c. Measurement of plasma vasopressin during hypertonic saline infusion
d. Therapeutic trial of desmopressin (1-desamino-8-D-arginine-vasopressin) treatment
e. Water deprivation test.

9

The Adrenal Glands

9.1 Hormones secreted in physiologically important amounts by the adrenal cortex in the male include
  a. Aldosterone
  b. Cortisol
  c. Noradrenaline
  d. Renin
  e. Testosterone.

9.2 Physiologically important roles of glucocorticoids include
  a. Inhibition of ACTH (adrenocorticotrophic hormone) secretion
  b. Maintenance of blood pressure
  c. Stimulation of hepatic gluconeogenesis
  d. Stimulation of protein synthesis
  e. Stimulation of renal sodium excretion.

9.3 Factors of physiological importance in the regulation of aldosterone secretion include
  a. ACTH (adrencorticotrophic hormone)
  b. Angiotensin II
  c. Blood pressure
  d. Plasma osmolality
  e. Renal blood flow.

9.4 Cortisol
a. Concentration in the plasma is normally highest in the morning
b. Is not essential for life
c. Is not normally excreted in the urine
d. In the plasma is approximately 50% bound to protein
e. Synthesis requires the enzyme, steroid 21-hydroxylase.

9.5 Recognized causes of adrenal cortical hypofunction include
a. Amyloidosis
b. Autoimmune disease
c. Hypoglycaemia
d. Pituitary tumour
e. Sodium depletion.

9.6 Recognized features of primary adrenal failure include
a. Hirsutism
b. Hyperkalaemia
c. Pallor of mucous membranes
d. Postural hypotension
e. Pre-renal uraemia.

9.7 Characteristic features of acute adrenal failure (Addisonian crisis) include
a. Alkalosis
b. Hyponatraemia
c. Hyperglycaemia
d. Natriuresis
e. Peripheral circulatory failure.

9.8 Tests of value in the diagnosis of suspected adrenal cortical failure include

a. Dexamethasone suppresssion test
b. Glucose tolerance test
c. Midnight plasma cortisol concentration
d. Short synacthen test
e. 24hr urine cortisol excretion.

9.9 In a patient suspected of having adrenal failure, the following findings would suggest an adrenal, rather than a pituitary cause:

a. Failure of plasma cortisol concentration to increase after administration of depot Synacthen daily for three days
b. Pigmentation of buccal mucosa
c. Low plasma cortisol concentration at 9.00am
d. Low plasma gonadotrophin concentrations
e. Presence of vitiligo.

9.10 Recognized causes of Cushing's syndrome (adrenal cortical hyperfunction) include

a. Carcinoid tumour
b. Haemochromatosis
c. Pituitary tumour
d. Phaeochromocytoma
e. Small (oat) cell carcinoma of bronchus.

9.11 Recognized features of Cushing's syndrome (adrenal cortical hyperfunction) include

a. Glucose intolerance
b. Hirsutism
c. Menstrual disturbance
d. Proximal myopathy
e. Presence of anti-adrenal autoantibodies.

39

9.12 Tests of value in the diagnosis of Cushing's syndrome (adrenal cortical hyperfunction) include
a. Dexamethasone suppression test
b. Glucose tolerance test
c. Insulin tolerance test
d. Measurement of plasma aldosterone concentration
e. Measurement of 24hr urine cortisol excretion.

9.13 In a patient with Cushing's syndrome, the following findings would suggest that a pituitary tumour was responsible
a. Elevated midnight plasma cortisol concentration
b. Hypertension
c. Low plasma ACTH concentration
d. Rapid progression of condition
e. Suppression of cortisol secretion in a high-dose dexamethasone suppression test.

9.14 Recognized features of Conn's syndrome include
a. Hypertension
b. Hypocalcaemia
c. Hypokalaemia
d. Muscle weakness
e. Polyuria.

9.15 Increased plasma aldosterone concentration is a recognized occurrence in patients with
a. Cushing's disease
b. Nephrotic syndrome
c. Phaeochromocytoma
d. Renal artery stenosis
e. Sheehan's syndrome.

9.16 Congenital adrenal hyperplasia
a. Can be causally associated with hypertension
b. Can present with precocious pseudopuberty in male infants
c. Is a cause of Cushing's syndrome
d. Is most frequently due to steroid 11-hydroxylase deficiency
e. Can present with a salt-losing state in the neonatal period.

9.17 Phaeochromocytomas
a. Are benign in the majority of cases
b. Are extra-adrenal in approximately 10% of cases
c. Are multiple in approximately 50% of cases
d. Cause an increase in the urinary excretion of 5-hydroxyindoleacetic acid (5HIAA)
e. Of the adrenals tend to secrete adrenaline as their major product.

10

The Thyroid Gland

10.1 The following statements concerning thyroid hormones are correct:

a. The plasma concentration of reverse triiodothyronine (rT3) tends to increase during starvation

b. The production of TSH (thyroid stimulating hormone) by the pituitary is independent of the hypothalamus

c. Thyroxine and triiodothyronine are secreted by the C cells of the thyroid

d. Thyroxine is the major product of the thyroid gland

e. Triiodothyronine is produced mainly by deiodination of thyroxine in skeletal muscle.

10.2 Thyroid hormone synthesis

a. Involves the iodination of tryptophan residues in thyroglobulin

b. Is inhibited by carbimazole and related drugs

c. Is stimulated by TSH (thyroid-stimulating hormone)

d. Requires the active uptake of iodide into the thyroid

e. Takes place in the thyroid follicular cells.

10.3 The release of thyroid hormones from the thyroid

a. Involves lysosomal degradation of colloid

b. Is accompanied by the release of a small amount of thyroglobulin into the plasma

c. Is increased in hyperthyroidism

d. Is independent of TSH (thyroid stimulating hormone)

e. Is inhibited by carbimazole and related drugs.

10.4 The following statements concerning the thyroid hormones in blood are correct:
a. The major transport protein for thyroxine in the blood is thyroglobulin
b. The normal free triiodothyronine concentration in plasma is approximately twice that of free thyroxine
c. The reference range for total triiodothyronine concentration in plasma is approximately 1–3nmol/l
d. Thyroid-binding proteins in the plasma are normally nearly 100% saturated with thyroxine
e. Thyroxine is protein-bound to a greater extent than triiododthyronine.

10.5 Decreased total concentrations of thyroid hormones in the plasma with normal concentrations of free thyroxine occur in
a. Nephrotic syndrome
b. Patients treated with androgens
c. Patients treated with corticosteroids
d. Patients treated with phenytoin
e. Pregnancy.

10.6 Plasma free thyroxine concentration
a. Can be normal in mild hypothyroidism
b. Has a reference range of approximately 60–150μmol/l
c. Is an important factor in controlling TSH (thyroid stimulating hormone) secretion by the pituitary
d. Is invariably elevated in hyperthyroidism
e. Is the most sensitive test for the diagnosis of hyperthyroidism.

10.7 The following statements about TSH (thyroid stimulating hormone) are correct:
   a. It is extensively protein-bound in plasma
   b. It is routinely measured in cord blood to screen for congenital hypothyroidism
   c. Its plasma concentration is frequently elevated in patients with pituitary tumours
   d. Its plasma concentration varies in parallel with plasma thyroxine concentration in most patients with thyroid disease
   e. The reference range in plasma is approximately 0.2–5.0mU/l.

10.8 The following statements about the response of plasma TSH (thyroid stimulating hormone) concentration to the injection of TRH (thyrotropin releasing hormone) are correct:
   a. A lack of response is diagnostic of hyperthyroidism
   b. A normal response excludes hypopituitarism
   c. It is characteristically prolonged in hypothalamic disease
   d. It is predictable from the basal (unstimulated) TSH concentration
   e. The peak effect is normally seen at one hour after the injection.

10.9 Autoantibodies have been described in man which recognize
   a. The TSH receptor
   b. Thyroglobulin
   c. Thyroid micosomes
   d. Thyroxine
   e. Triiodothyronine.

10.10 Recognized causes of hyperthyroidism include
  a. Graves disease
  b. Iodine deficiency
  c. Medullary carcinoma of thyroid
  d. Multinodular colloid goitre
  e. Treatment with lithium.

10.11 Recognized clinical features of hyperthyroidism include
  a. Carpal tunnel syndrome
  b. Heat intolerance
  c. Pericardial effusion
  d. Proximal myopathy
  e. Weight loss.

10.12 In hyperthyroidism due to intrinsic thyroid disease recognized laboratory findings include
  a. Hypercalcaemia
  b. Hypercholesterolaemia
  c. Hypoglycaemia
  d. Low plasma TSH (thyroid stimulating hormone) concentration
  e. Macrocytic anaemia.

10.13 Recognized causes of hypothyroidism include
  a. Excessive iodine intake
  b. Hashimoto's disease
  c. Hypopituitarism
  d. Thyroid agenesis
  e. Treatment of hyperthyroidism with radioactive iodine.

45

10.14 Recognized features of hypothyroidism include
  a. Bradycardia
  b. Dementia
  c. Diarrhoea
  d. Intolerance of cold
  e. Tremor.

10.15 The following statements are true of hypothyroidism:
a. Hyponatraemia is a recognized complication
b. In early disease, the plasma free triiodothyronine concentration is more likley to be low than the free thyroxine concentration
c. It is frequently the presenting feature of pituitary disease
d. It is usually associated with goitre
e. There is a recognized association with pernicious anaemia.

10.16 The combination of a low plasma TSH concentration and a high-normal free thyroxine concentration is compatible with
a. Early primary hypothyroidism
b. Inadequately treated hyperthyroidism
c. Low plasma thyroxine-binding globulin concentration in a euthyroid patient
d. Sub-clinical hyperthyroidism
e. Treated primary hypothyroidism.

10.17 Recognized actions of drugs in relation to thyroid function include
a. Displacement of thyroxine from thyroxine binding globulin by salicylates
b. Inhibition of thyroxine secretion by iodine
c. Stimulation of conversion of thyroxine to triiodothyronine by carbimazole
d. Stimulation of thyroxine binding globulin synthesis by corticosteroids
e. Stimulation of thyroxine binding globulin synthesis by oestrogens.

11

The Gonads

11.1 The following statements about the testes and their function are correct:
a. Inhibin is secreted by Sertoli cells
b. Leydig cell function is controlled by LH (luteinizing hormone)
c. Spermatogenesis is stimulated by FSH (follicle stimulating hormone)
d. Testicular function declines rapidly in the elderly
e. Testosterone is secreted by Sertoli cells.

11.2 Testosterone
a. Depends on metabolism to androstenedione for its physiological activity
b. Has a normal plasma concentration in the male of approximately the range 10-25nmol/l
c. Inhibits LH (luteinizing hormone) secretion
d. In the blood is more than 90% bound to sex hormone-binding globulin
e. Is secreted by the ovaries.

11.3 Oestradiol
a. Can be synthesized from testosterone in adipose tissue
b. In men is mainly produced by the testes
c. In plasma is bound mainly to sex hormone-binding globulin
d. Stimulates LH (luteinizing hormone) secretion immediately before ovulation
e. Stimulates the synthesis of sex hormone-binding globulin.

11.4 Sex hormone binding globulin

a. Concentration in plasma decreases in hyperthyoidism

b. Concentration in plasma is higher in males than in females

c. Concentration in plasma tends to be lower in obese individuals

d. Has greater affinity for testosterone than for oestradiol

e. Is the major transport protein for progesterone.

11.5 Hypogonadism in males

a. Can be associated with high plasma gonadotrophin concentrations

b. Can be diagnosed with confidence if a boy has not entered puberty by the age of 15 years

c. Can be due to a chromosomal disorder

d. Is unlikely to be of pituitary or hypothalamic origin if plasma testosterone concentration increases after the administration of clomiphene for seven days

e. Should be treated with testosterone to restore fertility.

11.6 Recognized causes of gynaecomastia include

a. Bronchial carcinoma

b. Chronic liver disease

c. Low plasma sex hormone-binding globulin concentration

d. Re-feeding after starvation

e. Treatment with spironolactone.

11.7 Recognized causes of amenorrhoea include

a. Chronic renal failure

b. Intensive exercise

c. Polycystic ovary syndrome

d. Pregnancy

e. Turner's syndrome.

11.8 The following statements about amenorrhoea are correct:
   a. About 50% of cases are related to hyperprolactinaemia
   b. It is classified as primary if it is due to an an abnormality of ovarian function
   c. It is likely to be due to ovarian failure if plasma FSH concentration is elevated
   d. It is very likely to occur if body weight falls below 90% of normal
   e. The progestogen challenge test can be used to diagnose failure of ovulation as a cause.

11.9 The following statements about hirsutism are correct:
   a. It can be due to a high plasma concentration of sex hormone-binding globulin
   b. It can be the presenting feature of polycystic ovary syndrome
   c. It can occur as a consequence of ovarian failure
   d. It is usually accompanied by menstrual irregularity
   e. The plasma testoterone concentration is usually clearly elevated.

11.10 Conditions which can cause virilization include
   a. Adrenal carcinoma
   b. Congenital adrenal hyperplasia
   c. Cushing's syndrome
   d. Ovarian carcinoma
   e. Turner's syndrome.

11.11 Typical findings in a patient with polycystic ovary syndrome include
a. FSH concentration in plasma much greater than that of LH
b. Hyperinsulinaemia
c. Oligomenorrhoea
d. Slightly elevated plasma testosterone concentration
e. Weight loss.

11.12 Infertility
a. Due to intrinsic gonadal failure is associated with low plasma concentrations of gonadotrophins
b. In the male is frequently due to endocrine disease
c. Is a recognized consequence of hyperprolactinaemia
d. Is classified as primary if it is due to gonadal disease
e. Is unlikely to be due to failure of ovulation if plasma progesterone is >30nmol/l on day 21 of the menstrual cycle.

11.13 Biochemical changes that occur in maternal plasma during normal pregnancy include increases in the concentration/activity of
a. Albumin
b. Alkaline phosphatase
c. Cortisol
d. Glucose (fasting)
e. Urea.

11.14 During pregnancy
a. A falling maternal plasma oestradiol concentration is a reliable indicator of fetal growth retardation
b. Fetal lung maturity can be assessed by measurement of lecithin in maternal plasma
c. The glomerular filtration rate increases
d. The peak concentration of human chorionic gonadotrophin (hCG) in maternal plasma is attained at about 12 weeks
e. The renal threshold for glucose increases.

12

Diabetes Mellitus

12.1 Insulin
a. Inhibits hepatic ketogenesis
b. Is secreted in response to a decrease in blood glucose concentration
c. Secretion is accompanied by equimolar secretion of C-peptide
d. Stimulates adipose tissue triglyceride synthesis
e. Stimulates hepatic glucose uptake.

12.2 Glucose
a. Concentration in plasma is usually 10–15% higher than in whole blood
b. Concentration in the cerebrospinal fluid is approximately 65% of the concentration in blood
c. In the urine is diagnostic of diabetes mellitus
d. Is synthesized in the liver in the fasting state
e. Metabolism by red blood cells in vitro can be prevented by collecting blood into a tube containing potassium oxalate

12.3 Insulin-dependent diabetes mellitus
a. Can cause growth retardation in children
b. Has many of the characteristics of an autoimmune disease
c. Is often associated with obesity
d. Shows a strong association with histocompatibility antigens DR3 and DR4
e. Usually presents at an older age than non-insulin-dependent diabetes.

12.4 Non-insulin-dependent diabetes

a. Can be associated with resistance to the action of insulin
b. Frequently presents with ketoacidosis
c. Is also known as Type 1 diabetes
d. Is rarely symptomatic
e. Occurs less frequently than insulin-dependent diabetes.

12.5 Presenting features of diabetes mellitus include

a. Abdominal pain
b. Asymptomatic glycosuria
c. Coma
d. Renal calculi
e. Weight loss.

12.6 Long-term sequelae of diabetes mellitus include

a. Cataracts
b. Neuropathy
c. Osteomalacia
d. Pancreatic insufficiency
e. Renal failure.

12.7 The following findings are diagnostic of diabetes mellitus in a patient with clinical features of the condition:

a. Capillary blood glucose concentration of >11.1mmol/l 2 hours after a meal
b. Fasting venous plasma glucose of 8mmol/l
c. Glycated haemoglobin (HbA$_1$) 7%
d. Ketonuria
e. Random venous plasma glucose of 14mmol/l.

12.8 The following statements are true of the glucose tolerance test:

a. A fasting capillary blood glucose concentration of 6mmol and 2 hour value of 10mmol/l indicates impaired glucose tolerance

b. It is a valuable test in the diagnosis of intestinal malabsorption

c. It is necessary for the diagnosis of diabetes mellitus in the majority of suspected cases

d. Patients should decrease their carbohydrate intake for 3 days before the test

e. The standard dose of glucose is 50grams.

12.9 Glycated haemoglobin (HbA_1)

a. Has decreased affinity for oxygen in comparison to normal haemoglobin (HbA)

b. Is a genetically determined variant haemoglobin

c. Is the major form of haemoglobin in patients with poorly controlled diabetes

d. Is not detectable in non-diabetic individuals

e. Measurements are usually expressed as a percentage of the total haemoglobin.

12.10 In assessing metabolic control in patients with diabetes mellitus

a. Measurements of plasma fructosamine may be helpful

b. The presence of glycosuria excludes recent hypoglycaemia

c. The presence of glycosuria in a patient with non-insulin-dependent diabetes indicates a need for treatment with insulin

d. The presence of ketonuria is always indicative of incipient ketoacidosis

e. The proportion of glycated haemoglobin (HbA_1) gives an indication of the mean blood glucose concentration over the previous 6–8 weeks.

12.11 Recognized laboratory findings in untreated diabetic ketoacidosis include
a. Elevated plasma β-hydroxybutyrate concentration
b. Elevated plasma urea concentration
c. Hypokalaemia
d. Increased plasma osmolality
e. Low arterial $P\text{CO}_2$.

12.12 Clinical features of diabetic ketoacidosis include
a. Abdominal pain
b. Bradycardia
c. Dehydration
d. Hyperventilation
e. Tetany.

12.13 Factors involved in the pathogenesis of diabetic ketoacidosis include
a. Decreased synthesis of malonyl-CoA
b. Decreased tissue glucose uptake
c. Increased hepatic glucose synthesis
d. Inhibition of glycolysis
e. Inhibition of lipolysis.

12.14 In the management of diabetic ketoacidosis
a. Hypotonic saline is the replacement fluid of choice
b. Insulin administration should be stopped once a normal blood glucose concentration has been attained
c. Insulin is best given subcutaneously
d. Intravenous bicarbonate should be given routinely
e. Potassium replacement is usually required even if the patient is hyperkalaemic at presentation.

12.15 In a patient with non-ketotic, hyperosmolar hyperglycaemia
a. Diabetes may not have been previously diagnosed
b. Long-term treatment with insulin will usually be required
c. Pre-renal uraemia is likely to be present
d. Significant acidosis is unlikely to be present
e. There is a risk of vascular thrombosis.

12.16 Recognized abnormalities of lipid metabolism associated with diabetes mellitus include
a. Decreased activity of lipoprotein lipase
b. Hypertriglyceridaemia
c. Increased adipose tissue lipolysis
d. Increased concentration of HDL (high density lipoprotein) cholesterol
e. Increased concentration of VLDL (very low density lipoprotein).

12.17 A positive test for reducing substances in the urine can occur in the absence of diabetes in
a. Fanconi syndrome
b. Galactosaemia
c. Individuals taking large doses of vitamin C
d. Normal pregnancy
e. Urine adulterated with sucrose.

13

Hypoglycaemia

13.1 The following hormones are correctly paired with their metabolic effects:
a. Adrenaline: stimulation of glycogenolysis
b. Cortisol: stimulation of gluconeogenesis
c. Glucagon: inhibition of gluconeogenesis
d. Growth hormone: stimulation of triglyceride synthesis
e. Insulin: stimulation of hepatic glucose uptake.

13.2 Clinical features of hypoglycaemia include
a. Ataxia
b. Polyuria
c. Sweating
d. Tremor
e. Weight loss.

13.3 Patients are more likely to experience symptoms of hypoglycaemia when their blood glucose concentration is low if
a. Cerebral blood flow is impaired
b. The blood glucose concentration has fallen rapidly
c. They are being treated with β-adrenergic antagonists
d. They are chronically hypoglycaemic
e. They are elderly.

13.4 Factors responsible for the maintenance of a normal blood glucose concentration during fasting include

a. Decreased glucagon secretion
b. Decreased insulin secretion
c. Decreased use of triglyceride as an energy substrate
d. Gluconeogenesis
e. Glycogenolysis.

13.5 Possible causes of hypoglycaemia in patients with insulin-dependent diabetes mellitus include

a. Alcohol ingestion
b. Co-existent hypopituitarism
c. Decreased food intake
d. Decreased physical activity
e. Inadvertent intravenous injection of insulin.

13.6 Fasting hypoglycaemia is a recognized complication of

a. Acromegaly
b. Adrenal failure
c. Glycogen storage disease type I (glucose-6-phosphatase deficiency)
d. Treatment with biguanides
e. Treatment with sulphonylureas.

13.7 Causes of primarily reactive hypoglycaemia include

a. Chronic pancreatitis
b. Hypopituitarism
c. Nesidioblastosis
d. Post-gastrectomy syndrome
e. Severe liver disease.

13.8 Factors implicated in hypoglycaemia related to alcohol include
a. Associated liver disease
b. Chronic pancreatic insufficiency
c. Impaired gluconeogenesis
d. Increased cortisol secretion
e. Increased insulin secretion in response to glucose ingestion.

13.9 Insulinomas
a. Are malignant in about 50% of cases
b. Are tumours of the exocrine pancreas
c. Are usually multiple
d. Can be a feature of multiple endocrine neoplasia
e. Characteristically cause fasting hypoglycaemia.

13.10 Procedures which are of value in the investigation of hypoglycaemia include the measurement of
a. 24hr urinary glucose excretion
b. Plasma C-peptide concentration during insulin-induced hypoglycaemia
c. Plasma glucagon concentration
d. Plasma β-hydroxybutyrate concentration
e. Plasma insulin concentration after an overnight fast.

13.11 The following conditions are recognized causes of hypoglycaemia in infants
a. Defects of fatty acid β-oxidation
b. Galactokinase deficiency
c. Islet cell hyperplasia
d. Maternal diabetes
e. Prematurity.

13.12 The following statements about hereditary fructose intolerance are correct:

a. Clinical manifestations are due to a high plasma fructose concentration

b. It is more likely to be present in breast-fed infants

c. It occurs only in males

d. Renal tubular dysfunction is a recognized complication

e. There is usually complete absence of the enzyme fructose-1-phosphate aldolase.

13.13 The following statements are true of ketotic hypoglycaemia:

a. Hypoglycaemia is thought to be due to decreased gluconeogenesis

b. It is a rare cause of hypoglycaemia in infancy

c. It is associated with an elevated plasma insulin concentration

d. It occurs more frequently in infants who are small for gestational age

e. Plasma alanine concentration is usually elevated.

14

Calcium, Phosphate, Magnesium and Bone

14.1 The following statements about calcium are correct:
a. It is approximately 50% bound to globulins in the plasma
b. Its absorption from the gut is under the direct control of parathyroid hormone
c. Most of the calcium filtered by the glomeruli is reabsorbed by the renal tubules
d. Most of the calcium in bone is freely exchangeable with the plasma
e. The reference range in plasma is approximately 2.3–2.6mmol/l.

14.2 In bone
a. Osteoblasts secrete alkaline phosphatase
b. Osteoclasts are responsible for bone formation
c. Osteoid is composed principally of polysaccharides
d. Resorption and new bone formation is a continuous process
e. The principal consituent of the mineral is calcium carbonate.

14.3 Parathyroid hormone
a. Increases renal tubular calcium reabsorption
b. Is phosphaturic
c. Is a steroid hormone
d. Is secreted in response to a fall in plasma calcium concentration
e. Stimulates the 25-hydroxylation of vitamin D.

14.4 Calcitriol (1,25-dihydroxy vitamin D)
  a. Acting alone, tends to decrease plasma phosphate concentration
  b. Inhibits its own formation
  c. Is secreted by the kidneys
  d. Stimulates calcium absorption from the gut
  e. Synthesis is stimulated by parathyroid hormone.

14.5 Calcitonin
  a. Deficiency leads to severe hypocalcaemia
  b. Is a polypeptide hormone
  c. Is present in the plasma in increased concentration during pregnancy
  d. Is secreted by thyroid follicular cells
  e. Stimulates osteoclasts.

14.6 Recognized causes of hypercalcaemia include
  a. Chronic pancreatitis
  b. Sarcoidosis
  c. Secondary hyperparathyroidism
  d. Thyrotoxicosis
  e. Treatment with thiazide diuretics.

14.7 Recognized clinical features of hypercalcaemia include
  a. Anorexia
  b. Calcification of basal ganglia
  c. Corneal calcification
  d. Muscle cramps
  e. Polyuria.

14.8 Recognized features of primary hyperparathyroidism include
a. Abdominal pain
b. Elevated plasma alkaline phosphatase activity
c. Goitre
d. Hypercalciuria
e. Hypophosphataemia.

14.9 Recognized causes of hypocalcaemia include
a. Paget's disease
b. Pseudohypoparathyroidism
c. Osteoporosis
d. Treatment with anticonvulsant drugs
e. Vitamin D deficiency.

14.10 Recognized features of hypocalcaemia include
a. Cataracts
b. Constipation
c. Convulsions
d. Hyperphosphataemia
e. Renal calculus formation.

14.11 Recognized causes of hypoparathyroidism include
a. Autoimmune disease
b. Di George syndrome
c. Hypermagnesaemia
d. Hyperphosphataemia
e. Vitamin D deficiency.

14.12 Recognized causes of hyperphosphataemia include
a. Fanconi syndrome
b. Intravenous infusion of glucose
c. Pseudohypoparathyroidism
d. Recovery phase of diabetic ketoacidosis
e. Renal failure.

14.13 The following statements about hypophosphataemia are true:

a. It can cause cause a decrease in 2,3-bisphospho-glycerate concentration in red blood cells

b. It is a feature of the tumour lysis syndrome

c. It is a recognized consequence of alcohol withdrawal

d. It is not usually of significance unless the concentration is less than 0.1mmol/l

e. Vitamin D deficiency may be responsible.

14.14 Magnesium

a. Deficiency can cause hypocalcaemia

b. Has a reference range in plasma of approximately 0.8–1.2mmol/l

c. Excretion is inhibited by parathyroid hormone

d. Excretion occurs mainly in the urine

e. In the body is mainly intracellular.

14.15 Causes of magnesium deficiency include

a. Alcoholic cirrhosis

b. Conn's syndrome

c. End stage renal failure

d. Malabsorption

e. Treatment with loop diuretics.

14.16 Recognized causes of osteoporosis include

a. Alcoholism

b. Hypophosphataemia

c. Ovarian failure

d. Prolonged immobilization

e. Vitamin D deficiency.

14.17 In osteoporosis
   a. Both the mineral and osteoid content of bone are decreased
   b. Hypocalcaemia is usual
   c. Plasma alkaline phosphatase activity is usually elevated
   d. Secondary hyperparathyroidism is usual
   e. Urinary hydroxyproline excretion is usually greatly increased.

15

Plasma Proteins

15.1 Known functions of proteins normally present in plasma include

a. Buffering

b. Catalysis

c. Enzyme inhibition

d. Maintenance of plasma oncotic pressure

e. Transport of metal ions.

15.2 When serum proteins are separated by electrophoresis on agarose gel

a. Absence of IgM causes an obvious decrease in the density of the γ-globulin band

b. Albumin migrates towards the cathode

c. An increase in IgG characteristically causes fusion of the β and γ bands

d. Fibrinogen forms a distinct band

e. The immunoglobulins remain near the origin.

15.3 Serum electrophoresis on a patient shows a decrease in albumin and γ-globulins, an increase in α_2- and β-globulins and a decrease in α_1-globulins. Possible causes include

a. Acute phase response

b. Hepatic cirrhosis

c. Nephrotic syndrome

d. Over-hydration

e. Protein-losing enteropathy.

15.4 Albumin
a. Concentration in the plasma is normally within the range 35–50g/l
b. Concentration in the plasma tends to fall in patients with septicaemia
c. Has a plasma half-life of approximately 8 days
d. Is the most abundant protein in plasma
e. Transports unconjugated bilirubin in the plasma.

15.5 α_1-Antitrypsin
a. Concentration in plasma typically increases in the nephrotic syndrome
b. Deficiency can be asymptomatic
c. Deficiency can be reliably diagnosed by electrophoresis of serum
d. Is an acute phase protein
e. Is a glycoprotein.

15.6 The following statements about α_1-antitrypsin deficiency are correct:
a. It is a cause of neonatal hepatitis
b. PiMZ heterozygotes have the same risk of developing lung disease as PiZZ homozygotes
c. The PiZZ genotype has a frequency of about 1 in three thousand in the United Kingdom
d. The protein is usually undetectable in the plasma in individuals with the PiZZ genotype
e. The risk of developing emphysema is greatly increased by cigarette smoking.

15.7 The following statements about plasma proteins are correct:
   a. A high total protein concentration in a patient with hypoalbuminaemia suggests an increase in the concentration of immunoglobulins
   b. Caeruloplasmin contains copper
   c. Haptoglobin concentration is characteristically increased in intravascular haemolysis
   d. α_2-Macroglobulin concentration tends to increase in protein-losing enteropathies
   e. Transferrin is an α_1-globulin.

15.8 Immunoglobulins
   a. Are classified according to the type of light chain that they contain
   b. Are found only in the plasma
   c. Are synthesized primarily in the liver
   d. Contain two heavy chains and two light chains
   e. Of the IgA class are secreted as dimers.

15.9 The following statements concerning immunoglobulins are correct:
   a. IgD has a major role in immediate hypersensitivity reactions
   b. IgG is normally present in the highest concentration in plasma
   c. IgG is the major antibody of the primary immune response
   d. IgM is pentameric
   e. IgM is the major antibody in bronchial mucus.

15.10 The following statements concerning immunoglobulins in infants' plasma are correct:
a. At birth, most of the IgG is maternal in origin
b. IgA concentration at birth is very low
c. IgG concentration falls immediately after birth
d. IgM concentration rises rapidly after birth
e. Normal adult concentrations of IgA are achieved by the end of the first year of life.

15.11 Recognized causes of hypogammaglobulinaemia include
a. Bruton's disease
b. Chronic lymphatic leukaemia
c. Hodgkins's disease
d. Recurrent bacterial infection
e. Systemic lupus erythematosus.

15.12 Hypergammaglobulinaemia is a recognized finding in
a. Patients treated with cytotoxic drugs
b. Nephrotic syndrome due to minimal change glomerulonephritis
c. Primary biliary cirrhosis
d. Rheumatoid disease
e. Severe malnutrition.

15.13 The following statements concerning paraproteins are correct:
a. When they occur in young people, they are usually benign
b. They are demonstrable by electrophoresis of serum in nearly all patients with myeloma
c. In myeloma, they are usually IgA
d. They are frequently associated with Raynaud's disease
e. They may be confused with fibrinogen if electrophoresis is performed on plasma.

15.14 Conditions associated with the presence of monoclonal paraproteins in serum include
a. Chronic infection
b. Franklin's disease
c. Multiple sclerosis
d. Rheumatoid arthritis
e. Waldenstrom's macroglobulinaemia.

15.15 Recognized complications of multiple myeloma include
a. Amyloidosis
b. Hyperuricaemia
c. Hypocalcaemia
d. Increased susceptibility to infection
e. Polycythaemia.

15.16 The following statements about cytokines are correct:
a. Colony-stimulating factors stimulate the growth of white blood cells
b. Interferons are anti-viral agents
c. Interleukins are inflammatory mediators
d. They mainly act at sites distant from their origin
e. Tumour necrosis factors stimulate the proliferation of T-lymphocytes.

16

Lipids and Lipoproteins

16.1 The following statements about cholesterol are correct:
a. Dietary fat is the body's only source of cholesterol
b. It is present in cell membranes
c. It is the major lipid in adipose tissue
d. It is the precursor of bile salts
e. It is the precursor of steroid hormones.

16.2 Chylomicrons
a. Are composed mainly of triglyceride
b. Are normally detectable in the plasma only in the fasting state
c. Are mainly removed from the circulation by the liver
d. Are the largest lipoprotein particles
e. Have a greater density than plasma.

16.3 The following statements about chylomicrons are correct:
a. Removal of triglyceride requires apoprotein CI as a cofactor
b. Their major apoprotein is B-100
c. Their metabolism requires the enzyme lipoprotein lipase
d. Their remnant particles acquire cholesterol ester from HDL
e. They transport dietary cholesterol.

16.4 The following statements about VLDL (very low density lipoproteins) are correct

   a. The triglyceride they contain is synthesized in the liver

   b. Their removal from the plasma depends on the activity of the enzyme lipoprotein lipase

   c. They are the precursors of LDL (low density lipoproteins)

   d. They contain apoprotein B-100

   e. They have approximately the same density as plasma.

16.5 VLDL (very low density lipoproteins)

   a. Are converted to IDL (intermediate density lipoprotein) by hepatic lipase

   b. Are not normally present in the plasma in the fasting state

   c. Are synthesized in the liver

   d. Have cholesterol as their major lipid component

   e. When present in excess in the plasma, cause hypertriglyceridaemia.

16.6 Low density lipoproteins (LDL)

   a. Are not normally detectable in plasma in the fasting state

   b. Are precursors of HDL (high density lipoproteins)

   c. Have cholesterol as their major lipid component

   d. Have B-100 as their main apoprotein

   e. Have β-mobility on electrophoresis of serum.

16.7 The following statements are true of low density lipoproteins (LDL):
a. A high plasma concentration is a risk factor for coronary heart disease
b. They are normally removed from the circulation by a receptor-dependent process
c. They are synthesized mainly in adipose tissue
d. Uptake into cells promotes intracellular cholesterol synthesis
e. When present in the plasma in excess, they can be removed by macrophages.

16.8 High density lipoproteins (HDL)
a. Act as a source of apoproteins for other lipoproteins
b. Are involved in reverse cholesterol transport
c. Are synthesized in liver and small intestine
d. Contain lipoprotein lipase
e. Do not cause the plasma to appear turbid when present in excess.

16.9 Factors positively associated with increased plasma HDL concentration include
a. Alcohol intake
b. Exercise
c. Genetic predisposition
d. Obesity
e. Plasma oestrogen concentration.

16.10 There is a recognized association between increased risk of coronary heart disease and
a. Decreased plasma apolipoprotein (a) concentration
b. Diabetes mellitus
c. Hypertension
d. Increased HDL cholesterol concentration
e. Smoking.

16.11 The following statements about lipids in blood are correct:
  a. After myocardial infarction, blood must be taken within 48 hours if representative results are to be obtained
  b. An HDL cholesterol concentration <0.9mmol/l is abnormally low
  c. The ideal cholesterol concentration is <6.5mmol/l
  d. The molar ratio HDL cholesterol/total cholesterol should ideally be >0.10
  e. Triglyceride concentration can be increased by recent food intake.

16.12 In the WHO classification of hyperlipidaemias
  a. Chylomicronaemia is a feature of type IV
  b. Increased LDL-cholesterol is a feature of type I
  c. Increased VLDL is a feature of type IIb
  d. In type IIa, the plasma triglyceride concentration is typically very high
  e. Type III is associated with the presence of a 'broad beta' band on plasma electrophoresis.

16.13 Recognized causes of hypercholesterolaemia include
  a. Alcohol ingestion
  b. Cholestatic liver disease
  c. Hypothyroidism
  d. Nephrotic syndrome
  e. Obesity.

16.14 Recognized features of familial hypercholesterolaemia include
  a. Decreased receptor-mediated removal of LDL from the plasma
  b. Eruptive xanthomata
  c. Higher prevalence in males than females
  d. Polygenic inheritance
  e. Tendon xanthomata.

16.15 Familial dysbetalipoproteinaemia (remnant hyper-lipoproteinaemia)

a. Causes increased plasma cholesterol and triglyceride concentrations

b. Is associated with a variant form of apolipoprotein E

c. Is associated with an increased risk of peripheral vascular disease

d. Is associated with the presence of fat deposits in the palmar skin creases

e. Occurs in approximately 1 in 100 individuals in the United Kingdom.

16.16 Chylomicronaemia in the fasting state

a. Can be due to deficiency of apolipoprotein B-48

b. Can be due to deficiency of the enzyme lipoprotein lipase

c. Is associated with a risk of pancreatitis

d. Can present in childhood

e. When genetically determined, is inherited as an autosomal dominant trait.

16.17 Familial combined hyperlipidaemia

a. Causes increased plasma HDL concentrations

b. Does not increase the risk of coronary heart disease

c. Is associated with increased synthesis of apolipoprotein B

d. Is inherited as an autosomal recessive trait

e. Usually causes a WHO type V hyperlipidaemia.

16.18 The following statements about the management of hyperlipoproteinaemia are correct:

a. Bile acid sequestrants decrease the conversion of cholesterol to bile acids

b. Bile acid sequestrants usually decrease plasma triglyceride concentrations significantly

c. Dietary advice should be given to all patients

d. Fibrates have no effect on plasma cholesterol concentration

e. Inhibitors of hydroxymethylglutaryl-CoA reductase (HMG CoA reductase) increase LDL receptor synthesis.

Clinical Enzymology

17.1 Causes of increased plasma enzyme activity include
- a. Cellular proliferation
- b. Damage to cell membranes
- c. Decreased excretion
- d. Enzyme induction
- e. Increased cell turnover.

17.2 Conditions in which plasma alkaline phosphatase activities greater than 5 times the upper limit of the reference range occur frequently include
- a. Inflammatory bowel disease
- b. Normal adolescence
- c. Normal pregnancy
- d. Primary biliary cirrhosis
- e. Primary hyperparathyroidism.

17.3 The following statements about isoenzymes are true:
- a. Acid phosphatase and alkaline phosphatase are isoenzymes
- b. Creatine kinase MB isoenzyme is present only in cardiac muscle
- c. Electrophoresis can be used to distinguish between isoenzymes of alkaline phosphatase
- d. Individual isoenzymes are always specific to one tissue
- e. The same isoenzyme of lactate dehydrogenase predominates in liver and skeletal muscle.

17.4 The following statements are true of acid phosphatase:

a. Elevated plasma activity is seen in Gaucher's disease

b. It is present in blood platelets

c. It is not present in bone

d. Its measurement in plasma provides a reliable test for distinguishing between benign and malignant prostatic disease

e. The prostatic isoenzyme is inhibited by tartrate.

17.5 The following statements about transaminases are correct:

a. Increased plasma activities occur mainly as a result of enzyme induction

b. In general, plasma alanine transaminase activity is elevated to a lesser extent than aspartate transaminase in non-hepatic disease

c. In hepatitis, an elevation of plasma enzyme activity often precedes the onset of jaundice

d. They are widely distributed in the body

e. Tissue-specific isoenzymes are readily distinguishable.

17.6 Aspartate transaminase activities more than ten times the upper limit of the reference range occur frequently in patients with

a. Acute hepatitis

b. Cholestatic liver disease

c. Chronic pancreatitis

d. Crush injuries

e. Myocardial infarction.

17.7 An increase in plasma γ-glutamyl transpeptidase activity, in the absence of changes in other plasma enzyme activities, is a recognized finding in patients with

a. A high alcohol intake
b. Acute hepatitis
c. Acute pancreatitis
d. Epilepsy treated with phenytoin
e. Myocardial infarction.

17.8 Elevated plasma hydroxybutyrate dehydrogenase activity is recognized to occur as a result of

a. Acute haemolysis
b. Acute hepatitis
c. Myocardial infarction
d. Myositis
e. Osteoporosis.

17.9 The following statements about creatine kinase are correct:

a. Brain and thyroid tissue contain the same isoenzyme
b. Forms of the enzyme found in plasma can contain one or both of two distinct monomers
c. Plasma enzyme activity often increases to more than five times the upper limit of the reference range following brain damage
d. The active enzyme is tetrameric
e. The enzyme normally present in plasma is mainly derived from skeletal muscle.

17.10 Recognized causes of a plasma creatine kinase activity greater than five times the upper limit of the reference range include

a. Grand mal convulsions
b. Hypothyroidism
c. Malignant hyperpyrexia
d. Rhabdomyolysis
e. Severe exercise.

17.11 Plasma amylase activities greater than five times the upper limit of the reference range are a recognized occurrence in patients with

a. Acute oliguric renal failure
b. Diabetic ketoacidosis
c. Intestinal obstruction
d. Perforated peptic ulcer
e. Trauma to skeletal muscle.

17.12 The following statements are true concerning the activities of enzymes in plasma following myocardial infarction:

a. Aspartate transaminase activity peaks at about 48h
b. Aspartate transaminase activity usually remains elevated for 4–5 days
c. Serial enzyme measurements may detect further infarction in the absence of electrocardiogrpahic changes
d. The maximum increase in creatine kinase activity occurs after about 24h
e. Total creatine kinase activity rises before a rise in the cardiac isoenzyme is detectable.

17.13 Diseases of bone in which plasma alkaline phosphatase activity is usually increased include

a. Multiple myeloma
b. Osteolytic metastatic disease
c. Paget's disease
d. Renal osteodystrophy
e. Rickets.

17.14 The following enzymes are correctly paired with conditions in which their measurement can be diagnostically useful:

a. Erythrocyte galactokinase activity: 'classical' galactosaemia

b. Erythrocyte glucose-6-phosphate dehydrogenase: glycogen storage disease type I

c. Plasma cholinesterase: persistent apnoea following administration of suxamethonium

d. Plasma creatine kinase: denervation of muscle

e. Plasma hydroxybutyrate dehydrogenase: myocardial infarction 4 days previously.

17.15 The following plasma enzymes are correctly paired with conditions in which they can be used to monitor progress or response to treatment:

a. β-N-acetylglucosaminidase: renal transplantation

b. Acid phosphatase: relief of biliary obstruction

c. Alkaline phosphatase: normal pregnancy

d. Angiotensin-converting enzyme: sarcoidosis

e. Creatine kinase: polymyositis.

18

Inherited Metabolic Diseases

18.1 Clinical manifestation of inherited metabolic disease can be the result of
a. Altered binding of a coenzyme to an enzyme
b. Decreased receptor synthesis
c. Decreased synthesis of a normal metabolite
d. Impairment of trans-membrane transport of a substance
e. Increased synthesis of a normal metabolite.

18.2 Glucose-6-phosphatase deficiency
a. Can be treated by continuous infusion of glucagon
b. Causes accumulation of glycogen in the liver
c. Is a cause of hypouricaemia
d. Is a cause of reactive hypoglycaemia
e. Results in impaired synthesis of glucose by gluconeogenesis.

18.3 Galactosaemia
a. Due to absence of the enzyme galactose-1-phosphate uridyl transferase is a harmless condition
b. Due to galactokinase deficiency usually presents with neonatal jaundice
c. Is a cause of a positive test for urinary reducing substances
d. Is a recognized cause of cataracts
e. Is treated by withdrawal of galactose and lactose from the diet.

18.4 Phenylketonuria

  a. Causes accumulation of toxic concentrations of tyrosine in the blood
  b. Is due to absence of the enzyme phenylalanine hydroxylase
  c. Is treated by total withdrawal of phenylalanine from the diet
  d. Occurs in approximately 1 in 10,000 infants born in the United Kingdom
  e. Requires treatment only in the neonatal period.

18.5 Steroid 21-hydroxylase deficiency

  a. Causes decreased synthesis of cortisol
  b. Causes an increased plasma concentration of 17-hydroxyprogesterone
  c. Is associated with decreased secretion of ACTH (adrenocorticotrophic hormone)
  d. Is associated with increased adrenal androgen synthesis
  e. Results in decreased formation of 11-deoxycortisol.

18.6 Cystic fibrosis

  a. Can present with intestinal obstruction
  b. Can present with recurrent respiratory infections
  c. Is a generalized disorder of endocrine secretion
  d. Is diagnosed by demonstrating a high sweat sodium or chloride concentration
  e. Occurs in about 1 in 10,000 babies in the United Kingdom.

18.7 The following statements about cystic fibrosis are correct:

a. In the United Kingdom, approximately 1 in 50 people are heterozygous for the defective gene

b. It can be screened for in the neontatal period by measurement of serum immunoreactive trypsin

c. It is most frequently due to a mutation designated ΔF_{508}

d. The product of the cystic fibrosis gene has been demonstrated to be a sodium pump

e. The prognosis is significantly improved if treatment is begun before symptoms develop.

18.8 The following inherited metabolic diseases are correctly paired with a causally related abnormality:

a. Galactosaemia: hypocalcaemia

b. Lipoprotein lipase deficiency: chylomicronaemia

c. Ornithine carbamoyl phosphate transferase deficiency: hyperammonaemia

d. Tyrosinaemia: jaundice

e. Wilson's disease: decreased urinary copper excretion.

18.9 The following statements are true concerning neonatal screening for phenylketonuria:

a. A positive test result is diagnostic of the condition

b. Screening is appropriate because early treatment improves the outcome

c. The test involves measuring the activity of the defective enzyme in red blood cells

d. The test is designed to have a sensitivity of as near 100% as possible

e. The test is performed on cord blood.

18.10 The following statements about pre-natal diagnosis of inherited disease are correct:
- **a.** Chorionic villus biopsy can be used to provide fetal tissue for analysis
- **b.** It is particularly applicable if an affected child has previously been born to the parents
- **c.** It should not be carried out before the second trimester of pregnancy
- **d.** Reliable diagnoses can often be made by analysis of maternal serum
- **e.** The test should be designed to avoid false positive results.

18.11 The following inherited metabolic diseases are correctly linked with their mode of inheritance:
- **a.** Acute intermittent porphyria: autosomal recessive
- **b.** Dubin-Johnson syndrome: autosomal recessive
- **c.** Duchenne muscular dystrophy: sex-linked recessive
- **d.** Familial hypercholesterolaemia: autosomal recessive
- **e.** Phenylketonuria: autosomal dominant.

18.12 The following inherited metabolic diseases are correctly paired with accepted forms of treatment:
- **a.** Congenital adrenal hyperplasia: ACTH (adrenocorticotrophic hormone)
- **b.** Cystinuria: penicillamine
- **c.** Galactosaemia: withdrawal of galactose-containing carbohydrates from the diet
- **d.** Homocystinuria: pyridoxal phosphate
- **e.** Wilson's disease: regular venesection.

18.13 The following statements about DNA analysis are correct:

a. A point mutation which affects a restriction site will decrease the size of the related restriction fragment

b. Analysis of DNA for diagnostic purposes is only useful if the product of the gene in question has been characterized

c. RFLP (restriction fragment length polymorphism) analysis is usually only useful for diagnosis if DNA from other members of the family can be analyzed

d. Southern blotting is a technique used to separate fragments of double-stranded DNA after enzymic digestion

e. The polymerase chain reaction (PCR) is used to produce DNA for genetic analysis from tiny amounts of starting material.

18.14 Restriction fragment length polymorphism (RFLP) analysis

a. For diagnostic purposes requires close linkage between a restriction site and the gene of interest

b. Involves electrophoretic separation of DNA fragments

c. Is a technique for measuring the sequence of bases in DNA

d. Requires an oligonucleotide probe that recognizes the restriction site

e. Requires knowledge of the base sequence of the gene of interest.

18.15 Gene probes

a. Can be prepared using messenger RNA to determine the sequence of bases

b. Can be used to detect mutant genes

c. Can be used to detect normal genes

d. Consist of double-stranded RNA

e. Must contain at least 100 bases.

19

Haemoproteins, Porphyrins and Iron

19.1 Normal adult haemoglobin
a. Binds carbon monoxide avidly
b. Contains four haem molecules
c. Contains iron in the iron III (Fe^{3+}) form
d. Contains two α and two β polypeptide chains
e. Is designated HbA.

19.2 Genetically determined abnormalities of haemoglobin synthesis can affect its
a. Concentration in the blood
b. Oxygen-carrying capacity
c. Solubility
d. Stability
e. Susceptibility to oxidation.

19.3 Methaemoglobin
a. Can be produced as a result of an inherited metabolic disorder of haem synthesis
b. Imparts a blue colour to the plasma
c. Is incapable of carrying oxygen
d. Is produced spontaneously in health
e. When produced as a result of exposure to drugs, is formed irreversibly.

19.4 In the reactions leading to the synthesis of haem

a. δ-aminolaevulinate dehydratase is responsible for the formation of porphobilinogen

b. Ferrochelatase is required for the incorporation of iron into protoporphyrin IX

c. Iron in the iron III (Fe^{3+}) form is incorporated into haem

d. Porphyrinogens are reduced to porphyrins

e. The synthesis of δ-aminolaevulinic acid is the first reaction unique to the pathway.

19.5 The following statements are true of the porphyrias:

a. Acute porphyrias are characterized mainly by photosensitivity

b. Cutaneous manifestations are due to excessive production of porphyrin precursors

c. Erythropoietic porphyrias present acutely

d. Most are inherited as autosomal dominant conditions

e. Variegate porphyria is the most prevalent type in the United Kingdom.

19.6 In acute intermittent porphyria

a. Attacks can be precipitated by alcohol

b. Blistering of the skin occurs

c. Muscle weakness can be the presenting feature

d. Patients are asymptomatic between attacks

e. The defective enzyme responsible for the condition is ferrochelatase.

19.7 The following statements are true of acute porphyrias:
a. Porphyrin precursors are excreted in the urine during attacks
b. Pregnancy can precipitate an attack
c. Presentation is usually before puberty
d. They are often precipitated by treatment with narcotic analgesics
e. They can present with abdominal pain.

19.8 Cutaneous hepatic porphyria (porphyria cutanea tarda)
a. Causes increased excretion of porphobilinogen in the urine
b. Is often associated with excessive alcohol ingestion
c. Is inherited in approximately 50% of cases
d. May respond to treatment by venesection
e. Typically presents with jaundice.

19.9 The following statements about iron are correct:
a. Absorption is decreased by dietary ascorbic acid
b. In the plasma is mainly protein-bound
c. It is better absorbed from the gut as iron III (Fe^{3+}) rather than iron II (Fe^{2+})
d. It is lost from the body mainly by urinary excretion
e. It is mainly absorbed in the ileum.

19.10 Plasma iron concentration
a. Falls early in iron deficiency
b. Is often decreased in patients with rheumatoid arthritis
c. Is a reliable measure of the body's iron stores
d. Shows considerable physiological fluctuation
e. Tends to be increased in patients with hepatitis.

19.11 The following statements about iron-binding proteins are correct:

a. A high plasma ferritin concentration is diagnostic of iron overload

b. A low plasma ferritin concentration is diagnostic of depletion of the body's iron stores

c. Each molecule of transferrin binds one iron ion

d. Increased saturation of transferrin with iron is characteristic of haemochromatosis

e. Transferrin is normally approximately one-half saturated with iron.

19.12 Hereditary (idiopathic) haemochromatosis

a. Is a disorder of hepatic iron metabolism

b. Is diagnosed more frequently in males than in females

c. Is inherited as an autosomal dominant condition

d. Is strongly linked with histocompatibility antigen HLA-B27

e. Characteristically causes iron accumulation in hepatic parenchymal cells.

19.13 Recognized complications of haemochromatosis include

a. Diabetes mellitus

b. Growth retardation

c. Heart disease

d. Hepatocellular carcinoma

e. Rheumatoid arthritis.

20

Hyperuricaemia and Gout

20.1 Uric acid
a. Excretion occurs exclusively through the kidneys
b. Formation is inhibited by allopurinol
c. In the plasma is in part derived from the diet
d. Is an end product of protein metabolism
e. Is present in the plasma mainly as the urate ion at physiological pH.

20.2 The following statements about purine nucleotide and uric acid metabolism are correct:
a. *De novo* nucleotide synthesis leads to the synthesis of inosine monophosphate (IMP)
b. Salvage pathways convert purine bases to their parent nucleotides
c. Xanthine oxidase converts adenine to uric acid
d. Decreased activity of phosphoribosylpyrophosphate synthetase causes increased synthesis of uric acid
e. Increased activity of salvage pathway enzymes causes increased synthesis of uric acid.

20.3 Urate concentrations in plasma
a. Are always elevated (in relation to the reference range) in patients with an acute attack of gout
b. Are related to the risk of developing gout
c. Show considerable variation between different ethnic groups
d. Tend to be higher in women than in men
e. Tend to decrease with age.

20.4 Recognized causes of hyperuricaemia include
  a. Carcinomatosis
  b. Glycogen storage disease type I
  c. Renal failure
  d. Treatment with cytotoxic drugs
  e. Xanthine oxidase deficiency.

20.5 The following statements about gout are correct:
  a. It is a recognized complication of the Fanconi syndrome
  b. It is usually due to a well-characterized metabolic defect
  c. Clasically it causes a monoarthritis affecting the first metatarsophalangeal joint
  d. The majority of people with hyperuricaemia will develop gout at some time
  e. There is a recognized association with hypertriglyceridaemia.

20.6 The following are of potential value in the treatment of patients with gout:
  a. Indomethacin
  b. Inhibitors of xanthine oxidase
  c. Low doses of aspirin
  d. Reduction in alcohol intake
  e. Thiazide diuretics.

20.7 Hypouricaemia
  a. Can occur in patients treated with probenecid
  b. Can occur in patients with severe liver disease
  c. Does not cause any recognised clinical syndrome
  d. Is a common finding
  e. Occurs in patients with congential xanthinuria.

20.8 Substances which can can become deposited in joints and cause arthritis include

a. Calcium pyrophosphate

b. Hydroxyapatite

c. Monosodium urate

d. Sodium chloride

e. Urea.

21

Metabolic Aspects of Malignant Disease

21.1 Possible causes of an endocrine syndrome in a patient with cancer include
a. Coexistent endocrine disease
b. Destruction of normal endocrine tissue by the tumour
c. Ectopic hormone secretion by the tumour
d. Eutopic hormone secretion by the tumour
e. Production of a hormone-like substance by the tumour.

21.2 Recognized consequences of hormone secretion by tumours include
a. Hypothyroidism
b. Gynaecomastia
c. Acromegaly
d. Diabetes mellitus
e. Hypertension.

21.3 In Cushing's syndrome due to ectopic ACTH secretion by a small cell carcinoma of bronchus
a. Cortisol secretion can usually be suppressed by giving dexamethasone
b. Glycosuria is rarely present.
c. Hyperkalaemia is usual
d. Plasma ACTH concentration is usually only slightly elevated
e. The classic somatic manifestations of Cushing's may be absent.

21.4 The following tumours are correctly paired with a condition caused by hormone secretion by the tumour:

a. Carcinoid tumours: Cushing's syndrome

b. Carcinomas of breast: hypocalcaemia

c. Mesenchymal tumours: hyperglycaemia

d. Squamous cell carcinomas of bronchus: dilutional hyponatraemia

e. Uterine fibromyomata: polycythaemia.

21.5 The following tumours are frequently associated with ectopic antidiuretic hormone (ADH) secretion:

a. Adenocarcinoma of pancreas

b. Carcinoid tumour

c. Myeloma

d. Renal cell carcinoma

e. Small cell carcinoma of lung.

21.6 Factors which are believed to contribute to hypercalcaemia in patients with malignant disease include

a. Destruction of bone by metastatic deposits

b. Increased renal tubular reabsorption of calcium

c. Secretion of calcitonin by tumour cells

d. Secretion of substances resembling parathyroid hormone by tumour cells

e. Stimulation of osteoblasts by tumour-derived prostaglandins.

21.7 Massive destruction of tumour cells by cytotoxic drugs can cause

a. Acute renal failure

b. Hypokalaemia

c. Hypophosphataemia

d. Hyperuricaemia

e. Increased urinary urea excretion.

21.8 Metabolic derangements in patients with malignant disease

    a. Are always due to hormone secretion by the tumour

    b. Are only seen if metastases are present

    c. Are rarely of clinical significance

    d. Are uncommon except with small cell carcinomas of of the bronchus

    e. May occur as a consequence of treatment.

21.9 Factors involved in the pathogenesis of cachexia in patients with malignant disease include

    a. Abnormal metabolism by the tumour

    b. Intestinal obstruction

    c. Secretion of tumour necrosis factor

    d. Treatment with cytotoxic drugs

    e. Weight loss.

21.10 The following statements apply to carcinoid tumours:

    a. Primary tumours occur only in the gut

    b. They are highly malignant

    c. They are often asymptomatic

    d. They occur most frequently in appendix and ileocaecal region

    e. When present in the gut, cause carcinoid syndrome only when there is metastasis to liver.

21.11 Recognized products of carcinoid tumours include

    a. 5-Hydroxytryptophan

    b. Peptide hormones

    c. Serotonin

    d. Steroid hormones

    e. Tryptophan.

21.12 The following statements about the carcinoid syndrome are correct
- **a.** A high urinary excretion of 5-hydroxyindoleacetic acid is characteristic
- **b.** Clinical features of nicotinic acid deficiency are a recognized association
- **c.** It may be caused by bronchial tumours in the absence of metastases
- **d.** It occurs in the majority of patients who have carcinoid tumours
- **e.** Symptoms may be provoked by alcohol ingestion.

21.13 The following statements about multiple endocrine neoplasia are correct
- **a.** Adrenal medullary tumours are a feature of types IIa and IIb
- **b.** Inheritance is autosomal dominant
- **c.** Medullary cell carcinoma of thyroid only occurs in type I
- **d.** Parathyroid adenomas are a feature of both type I and IIa
- **e.** The manifestations are constant in affected members of the same family.

21.14 Alpha-fetoprotein concentration in plasma
- **a.** At the time of diagnosis in patients with testicular teratomas is prognostically significant
- **b.** Can be measured as an effective screening test for primary liver cancer in the general population
- **c.** Increases during recovery from hepatitis
- **d.** Increases in the mother during early pregnancy
- **e.** Is elevated in approximately 90% of patients with malignant testicular teratomas.

21. Metabolic Aspects of Malignant Disease

21.15 Tumour markers which can be used reliably for screening for cancer in susceptible individuals include
a. Alpha-fetoprotein
b. Beta-human chorionic gonadotrophin
c. CA-125
d. Calcitonin
e. Prostatic acid phosphatase.

21.16 The following tumour markers are correctly paired with a tumour by which they are secreted:
a. Beta-human chorionic gonadotrophin (β-hCG): teratoma of testis
b. CA 15-3: carcinoma of breast
c. CA 19-9: adenocarcinoma of pancreas
d. Neuron-specific enolase: small cell carcinoma of bronchus
e. Placental alkaline phosphatase: carcinoma of prostate.

21.17 The following statements concerning tumour markers are true:
a. Carcinoembryonic antigen (CEA) concentration in plasma correlates poorly with with the size of colonic carcinomas
b. Choriocarcinomas secrete only the β-chain of human chorionic gonadotrophin (hCG)
c. In myeloma, the serum paraprotein concentration at the time of diagnosis is of prognostic significance
d. Persistence of α-fetoprotein concentration within the reference range after treatment of testicular teratoma excludes recurrence of the tumour
e. The presence of prostate-specific antigen in plasma is diagnostic of prostatic cancer.

97

Therapeutic Drug Monitoring and Chemical Aspects of Toxicology

22.1 Monitoring of plasma concentrations of a therapeutic drug is of potential value when it

a. Has a low therapeutic ratio

b. Has toxic effects which mimic the condition for which it is prescribed

c. Is excreted unchanged by the kidneys

d. Is extensively protein-bound

e. Is itself inactive, but is metabolized to a pharmacologially active metabolite.

22.2 Factors which can affect the concentration of a drug in the plasma in relation to the dose prescribed include

a. Compliance with treatment

b. Its rate of excretion

c. Its solubility in fat

d. The presence of physiological antagonists

e. The sensitivity of target cells/tissue to the drug.

22.3 For drugs in plasma, the therapeutic range is

a. Dependent on an assessment of clinical response

b. Dependent on renal function

c. Independent of the presence of other disease

d. The range of concentrations within which no interactions with other drugs is seen

e. The range of concentrations within which the drug is effective yet not toxic.

22.4 Measurement of plasma concentrations of the following drugs is of accepted value in determining dosage

a. Aminophylline

b. Cyclosporin A

c. Lithium

d. Insulin

e. Warfarin.

22.5 Monitoring plasma concentrations of phenytoin is clinically useful because

a. The drug has a narrow therapeutic ratio

b. The presence of a low concentration in a patient who is fit-free is an indication for withdrawing the drug

c. The relationship between plasma concentration and efficacy is unpredictable

d. There is considerable inter-individual variation between dose and steady-state plasma concentration

e. They may be affected by concomitant administration of other drugs.

22.6 For digoxin

a. Blood for measurement of plasma concentration should be drawn between 6 and 12 hours after the last dose

b. Measurement of plasma concentration may be used to assess compliance

c. Plasma concentrations in relation to dose are likely to be lower in hypothyroid patients

d. Sensitivity to the drug is increased by hyperkalaemia

e. There is a strong correlation between plasma concentration and therapeutic effect.

22.7 Lithium
a. Concentration in plasma reaches a steady state within 2 days after a change in dosage
b. Concentration in plasma should be measured 12 hours after the last dose
c. Excretion is potentiated by many diuretics
d. Is excreted by the kidneys
e. Is nephrotoxic.

22.8 Aminoglycoside antibiotics (e.g., gentamicin)
a. Are nephrotoxic
b. Have a long plasma half-life
c. Therapy is controlled by manipulation of both the amount of drug given and the interval between doses
d. Toxicity correlates better with peak, rather than trough, plasma concentration
e. Trough plasma concentrations should be measured 1 hour before the next dose is due.

22.9 Paracetamol
a. In overdose can cause renal failure
b. Is believed to be involved in the pathogenesis of Reye's syndrome
c. Toxicity can be reliably predicted from the plasma concentration measured within 4 hours of ingestion of the drug
d. Toxicity is due to a metabolite which is not produced at therapeutic doses
e. Toxicity usually becomes clinically obvious within 6 hours of an overdose.

22.10 In paracetamol poisoning
   a. Hypoglycaemia is a recognized complication
   b. If treatment with N-acetyl cysteine is indicated, it should ideally be given within 12 hours of the drug being ingested
   c. Jaundice is an early clinical sign
   d. The prothrombin time usually becomes prolonged within 8 hours of ingestion
   e. There is an early decrease in plasma albumin concentration.

22.11 Recognized clinical features of salicylate poisoning include
   a. Cyanosis
   b. Hyperventilation
   c. Irritability
   d. Sweating
   e. Tinnitus.

22.12 Recognized metabolic features of salicylate poisoning include
   a. Hyperlactataemia
   b. Hypernatraemia
   c. Hypoglycaemia
   d. Ketosis
   e. Non-respiratory acidosis.

22.13 Forced alkaline diuresis for salicylate poisoning
   a. Carries a risk of hypokalaemia
   b. Is unlikely to be effective unless urine pH exceeds 9.0
   c. Should not be attempted if the the patient has a systemic alkalosis
   d. Is usually required in adults if plasma salicylate concentration exceeds 2.0mmol/l
   e. Increases salicylate excretion by enhancing its tubular secretion.

22.14 Recognized features of lead poisoning include
   a. Abdominal pain
   b. Encephalopathy
   c. Gastrointestinal haemorrhage
   d. Increased red erythrocyte protoporphyrin
      concentration
   e. Increased urinary δ-aminolaevulinic acid excretion.

22.15 Abnormalities frequently found as a consequence of chronic excessive alchohol ingestion include
   a. Elevated plasma creatine kinase activity
   b. Elevated plasma γ-glutamyl transpeptidase activity
   c. Hyperglycaemia
   d. Hypertriglyceridaemia
   e. Increased mean red cell volume.

22.16 Measurements of drug/poison concentrations are of value in the management of patients poisoned with
   a. Carbon monoxide
   b. Cyanide
   c. Iron
   d. Lead
   e. Lithium.

22.17 Disturbance of acid–base homoeostasis is a recognized complication of poisoning with
   a. Barbiturates
   b. Ethanol
   c. Ethylene glycol
   d. Lead
   e. Paracetamol.

23

Clinical Nutrition

23.1 The following vitamins are correctly paired with their functions:
a. Ascorbic acid: essential for collagen synthesis
b. Folic acid: coenzyme for nucleic acid synthesis
c. Riboflavin: coenzyme for metabolism of pyruvate to acetyl-CoA
d. Thiamin: component of NAD
e. Vitamin E: antioxidant activity.

23.2 The following vitamins and deficiency diseases are correctly paired:
a. Nicotinic acid: beri-beri
b. Thiamin: Wernicke's encephalopathy.
c. Vitamin A: night blindness
d. Vitamin D: impaired blood coagulation
e. Vitamin B_{12}: megaloblastic anaemia.

23.3 Recognized causes of vitamin deficiency states include
a. Impaired metabolism
b. Inadequate intake
c. Increased losses
d. Increased requirements
e. Malabsorption.

23.4 Thiamin

a. Deficiency can present with peripheral neuropathy
b. Deficiency can cause memory loss
c. Deficiency is a recognized complication of carcinoid syndrome
d. Status can be assessed by measurement of red cell glutathione reductase
e. Stores in the liver are normally equivalent to six months requirements for the vitamin.

23.5 Vitamin D

a. Deficiency is an important cause of osteoporosis
b. Has little intrinsic activity in calcium homoeostasis
c. Is present in large quantities in green vegetables
d. Is synthesized in the skin
e. Undergoes 1-hydroxylation in the liver.

23.6 Elements known to be essential nutrients in man include

a. Fluorine
b. Iodine
c. Manganese
d. Molybdenum
e. Silicon.

23.7 The following statements about zinc are correct:

a. Deficiency is associated with poor wound healing
b. It is essential for the activity of red cell superoxide dismutase
c. Low plasma concentrations are diagnostic of zinc deficiency
d. Plasma concentrations increase significantly after eating
e. Urinary excretion increases following trauma.

23.8 Copper
a. Deficiency is frequently seen in patients receiving total parenteral nutrition
b. Deposition in tissues is a characteristic feature of wilson's disease
c. Is essential for the activity of cytochrome oxidase
d. Is present in caeruloplasmin
e. Is present in vitamin B_{12}.

23.9 The following statements about trace nutrients are true
a. Chromium deficiency predisposes to hypoglycaemia
b. Molybdenum is a cofactor for xanthine oxidase
c. Plasma concentrations are reliable indicators of their body stores
d. Selenium is a component of vitamin E
e. Selenium status can be assessed by measurement of red cell glutathione peroxidase activity.

23.10 Recognized metabolic complications of total parenteral nutrition include
a. Cholestasis
b. Hyperglycaemia
c. Hyperkalaemia
d. Hypoglycaemia
e. Hypophosphataemia.

23.11 During total parenteral nutrition
a. A rapid increase in plasma albumin concentration indicates that the patient is in positive nitrogen balance
b. Frequent measurements of plasma water-soluble vitamin concentrations are required
c. Hyponatraemia indicates a need to increase sodium input
d. Potassium input exceeding output suggests that the patient is catabolic
e. The plasma should be inspected for lipaemia.

105

24

Clinical Chemistry at the Extremes of Age

24.1 The plasma concentrations of analytes tend to vary as shown in healthy elderly people in relation to young adults

a. Albumin concentration: decreased

b. Alkaline phosphatase activity: increased

c. Cholesterol concentration: decreased

d. Creatinine concentration: decreased

e. Urate concentration: decreased.

24.2 Plasma analytes having significantly different reference ranges in neonates than in older children include

a. Aspartate transaminase

b. Calcium

c. Glucose

d. Phosphate

e. Potassium.

24.3 Causes of hypocalcaemia in infancy include

a. Di George syndrome

b. Exchange blood transfusion

c. High phosphate intake

d. Maternal vitamin D deficiency

e. Secondary hyperparathyroidism.

24.4 The following statements about 'physiological' neonatal jaundice are correct:
a. Enterohepatic circulation of bilirubin is in part responsible
b. It develops within a few hours of birth
c. It is due in part to immaturity of hepatic enzymes
d. It is due only to conjugated bilirubin
e. The plasma bilirubin concentration is usually less than 100μmol/l.

24.5 Causes of unconjugated hyperbilirubinaemia developing in the neonatal period include
a. Congenital cytomegalovirus infection
b. Crigler-Najjar syndrome
c. Galactosaemia
d. Rhesus blood group incompatibility
e. Urinary tract infection.

24.6 Causes of conjugated hyperbilirubinaemia developing in the neonatal period include
a. α_1-antitrypsin deficiency.
b. Biliary atresia
c. Breast milk jaundice
d. Hypothyroidism
e. Red cell pyruvate kinase deficiency

24.7 Tests of value in screening for metabolic causes of illness in the newborn include
a. Chromatography of urine for amino acids
b. Examination of urine for porphyrin precursors
c. Examination of urine for reducing substances
d. Measurement of arterial hydrogen ion concentration
e. Measurement of blood glucose concentration.

24.8 Recognized causes of hyperammonaemia in infancy include
a. Gilbert's syndrome
b. Inherited disorders of urea synthesis
c. Parenteral nutrition
d. Reye's syndrome
e. Severe infection.

24.9 Causes of true precocious puberty include
a. Adrenal tumours
b. Congenital adrenal hyperplasia
c. Ovarian tumours
d. Pineal tumours
e. Testicular tumours.

24.10 Recognized causes of delayed puberty include
a. Chronic renal failure
b. Coeliac disease
c. Hypothyroidism
d. Pituitary insufficiency
e. Treatment with corticosteroids.

24.11 The following statements about conditions associated with abnormal sexual development are correct:
a. Adrenal tumours are a cause of male pseudohermaphroditism
b. Congenital adrenal hyperplasia is a cause of virilization in females
c. Klinefelter's syndrome is a recognized cause of gynaecomastia
d. Steroid 5α-reductase deficiency is a cause of virilization in females
e. The karyotype in Turner's syndrome is 47XXY.

25

Case Histories

In this section, answers are given on the page following each set of questions. Adult reference ranges are given on pages 141-143.

25.1 A fit, elderly man has biochemical tests performed as part of a 'well-man' screen. The only abnormality is a serum alkaline phosphatase activity of 200iu/l. Possible causes include

a. Benign prostatic hypertrophy

b. Osteomalacia

c. Osteoporosis

d. Paget's disease of bone

e. Tumour metastasis to the liver.

25.2 An overnight dexamethasone suppression test is performed. Serum cortisol concentration at 0900h next morning is 180nmol/l. This result is compatible with

a. Alcoholism

b. Cushing's disease

c. Depression

d. Ectopic secretion of ACTH

e. Severe stress.

25.3 A plasma sample is observed to be lipaemic. Possible causes include

a. Chylomicronaemia

b. Increased HDL cholesterol concentration

c. Increased LDL cholesterol concentration

d. Type IIa hyperlipoproteinaemia

e. Uncontrolled diabetes mellitus.

25.1 b, d, e.
Plasma alkaline phosphatase is not elevated in
osteoporosis unless there is pathological fracture or
concomitant osteomalacia. The prostate contains
acid, not alkaline, phosphatase, but plasma activity is
not usually elevated in benign prostatic hypertrophy.

25.2 a, b, c, d, e.
Dexamethasone normally suppresses ACTH, and
hence cortisol, secretion. A failure of suppression is
characteristic of Cushing's syndrome whatever the
cause but can occur in other conditions too.

25.3 a, e.
Only chylomicrons and VLDL are sufficiently large to
scatter light and hence cause plasma to appear
lipaemic. Both can be present in uncontrolled
diabetes. Only LDL is increased in type IIa
hyperlipoproteinaemia.

25.4 An elderly woman complains of back pain: serum total protein concentration 85g/l; albumin, 30g/l. The presence of the following conditions could explain these abnormalities

a. Chronic osteomyelitis

b. Multiple myeloma

c. Osteoarthritis

d. Paget's disease of bone

e. Renal osteodystrophy.

25.5 An elderly man presents with an acute confusional state. His serum sodium concentration is 108mmol/l. Which of the following would suggest that this is due to water overload rather than to sodium depletion?

a. Low blood pressure

b. Plasma osmolality 230mmol/kg

c. Serum albumin concentration 48g/l

d. Serum urea concentration 3.2mmol/l

e. Urine sodium concentration 5mmol/l.

25.6 A middle-aged woman with a long history of rheumatoid disease complains of fainting episodes. Postural hypotension is demonstrable. Plasma sodium concentration is 128mmol/l. The sodium concentration of a random urine sample is 80mmol/l. The following diagnoses are compatible with these findings

a. Adrenal failure

b. Analgesic nephropathy

c. Syndrome of inappropriate antidiuresis

d. Over-treatment with diuretics

e. Purgative abuse.

25.4 a, b.

Osteomyelitis can cause a polyclonal increase in γ-globulins and myeloma a monoclonal increase. These changes are not seen with the other conditions. Both malignancy and chronic infection are frequently associated with low albumin concentration.

25.5 d.

Hypotension, low urine sodium excretion and increased albumin concentration are features of sodium depletion but this would be expected to cause an increase in urea (pre-renal uraemia). The osmolality is appropriate for the sodium concentration and does not differentiate between these two possible causes of hyponatraemia.

25.6 a, b, d.

Hyponatraemia with hypotension suggests sodium depletion (decreased ECF volume). Renal sodium excretion is inappropriately high, suggesting that renal sodium loss is responsible. Purgative abuse causes sodium loss from the gut. The syndrome of inappropriate antidiuresis causes hyponatraemia but not a decrease in ECF volume.

25.7 An elderly man is admitted to hospital with retention of urine. Plasma urea concentration is 48mmol/l; plasma creatinine, 520μmol/l. Which of the following additional findings would suggest that he might have underlying chronic renal failure?

a. Anaemia

b. High plasma alkaline phosphatase activity

c. Hyperphosphataemia

d. Hyponatraemia

e. Small kidneys on ultrasound examination.

25.8 A young man is admitted to hospital unconscious, having been knocked down by a motor car late one night. Plasma sodium concentration is 140mmol/l, potassium 4.2mmol/l, bicarbonate 25mmol/l, urea 6.2mmol/l, glucose 7.1mmol/l, osmolality 322mmol/kg. Likely diagnoses include

a. Acute adrenal insufficiency

b. Ethanol intoxication

c. Diabetes insipidus due to head injury

d. Salicylate poisoning

e. Water intoxication.

25.9 A middle-aged man presents with renal colic. The following findings would suggest a possible cause:

a. Co-existent psoriasis

b. Eruptive xanthomata

c. Evidence of malabsorption

d. History of acute arthritis affecting the big toe one year previously

e. Unconjugated hyperbilirubinaemia.

25.7 a, b, e.
Hyperphosphataemia and hyponatraemia occur in both acute and chronic renal failure. The kidneys are usually small in chronic renal failure (unless due to polycystic disease or amyloid). Patients with chronic renal failure are usually anaemic (decreased erythropoietin synthesis) and have renal osteodystrophy, causing elevated alkaline phosphatase.

25.8 b.
There is an 'osmolar gap' indicating the presence of an unmeasured solute in plasma at a concentration of approximately 30mmol/l. This could be alcohol. An 'osmolar gap' would not be expected in the other conditions.

25.9 a, c, d.
Hypercholesterolaemia and hyperbilirubinaemia do not cause renal calculi. Psoriasis is a recognized cause of hyperuricaemia, itself a cause of renal calculi. Malabsorption can cause increased urinary oxalate excretion and predispose to calculus formation.

25.10 An elderly lady complains of muscle weakness and
constipation. Plasma potassium concentration is
2.4mmol/l, bicarbonate 42mmol/l; urine potassium
excretion is 50mmol/24hr. Possible diagnoses
include

   a. Acute renal failure

   b. Conn's syndrome

   c. Purgative abuse

   d. Renal tubular acidosis

   e. Treatment with thiazide diuretics.

25.11 A baby born at term develops slight jaundice 48hrs
after birth. This becomes more severe, and his
nappy shows dark staining from urine although the
stool appears pale. Possible diagnoses include

   a. ABO blood group incompatibility

   b. Biliary atresia

   c. Crigler–Najjar syndrome

   d. Hypothyroidism

   e. Neonatal hepatitis.

25.12 A child of 6 is investigated for short stature. The
results of the following investigations could
suggest a possible cause:

   a. Measurement of plasma growth hormone
concentration during a glucose tolerance test

   b. Plasma creatinine concentration

   c. Plasma TSH concentration

   d. Sweat sodium concentration

   e. Testing urine for reducing substances.

25.10 b, e.
The high bicarbonate suggests that the hypokalaemia is associated with potassium depletion, and the urine potassium suggests that this is due to a renal cause. In purgative abuse, potassium is lost from the gut. Acute renal failure cause hyperkalaemia. Renal tubular acidosis can cause hypokalaemia but there is a systemic acidosis, and bicarbonate concentration is low.

25.11 b, e.
The dark urine and pale stool indicate conjugated hyperbilirubinaemia. In ABO blood group incompatibility, Crigler–Najjar syndrome and hypothyroidism, jaundice is due to unconjugated bilirubin.

25.12 b, c, d, e.
Growth hormone secretion is normally suppressed during a glucose tolerance test; this is used to diagnose excessive growth hormone secretion. The other tests might identify renal failure, hypothyroidism, cystic fibrosis and diabetes, respectively, all of which can cause growth retardation.

25.13 A patient with hypertension has a plasma potassium concentration of 2.8mmol/l; plasma bicarbonate is 34mmol/l. Possible diagnoses/ explanations which explain all these findings include

a. Chronic ingestion of liquorice
b. Conn's syndrome
c. Renal artery stenosis
d. Renal tubular acidosis with incidental hypertension
e. Treatment of essential hypertension with thiazide diuretic.

25.14 Four days after surgery for gall stones, a patient is noticed to be jaundiced. Serum bilirubin concentration is 90μmol/l, alkaline phosphatase activity 165IU/l; other 'liver function' tests are normal. Possible explanations include

a. Coincidental infectious hepatitis
b. Damage to common bile duct
c. Drug-induced cholestasis
d. Intraabdominal sepsis
e. Resorption of haematoma.

25.15 An elderly patient with atrial fibrillation and congestive cardiac failure is symptom-free on treatment with digoxin and a thiazide diuretic. Serum digoxin concentration, 10 hours after the previous dose, is 2.9nmol/l. The following statements are true:

a. Plasma potassium concentration should be measured
b. The concentration should be checked 3 hours after the next dose
c. The drug is probably unecessary and could be withdrawn without detriment to the patient
d. The patient is probably non-compliant
e. There is a risk of imminent renal failure.

25.13 b, c, e.
Hypokalaemia with a raised bicarbonate indicates
potassium depletion. This occurs in Conn's syndrome,
(primary aldosteronism) and in renal artery stenosis
(secondary aldosteronism). Hypertension occurs with
both conditions. Thiazides are used to treat
hypertension and can cause potassium depletion.
Liquorice ingestion can cause potassium wasting but
not hypertension. Plasma bicarbonate is decreased in
renal tubular acidoses.

25.14 b, c, d.
Post-operative jaundice can have many causes. Here,
the elevated alkaline phosphatase suggests
cholestasis, compatible with damage to the bile duct or
sepsis. The normal transaminase excludes hepatitis.
Resorption of a haematoma may cause conjugated
hyperbilirubinaemia, but not cholestatic jaundice.

25.15 a.
The blood has been drawn at an appropriate time after
the last dose. The concentration is slightly above the
therapeutic range; digoxin toxicity is potentiated by
hypokalaemia. Renal failure is not a feature of digoxin
toxicity; the patient is clearly taking the drug and there
is no reason to suspect that it is not required.

25.16 A young woman takes an overdose of paracetamol. She is discovered and admitted to hospital, approximately 36 hours after the overdose. Which of the following findings at this time are unlikely to be due to the paracetamol alone?

a. Arterial P_{CO_2} 7.8kPa

b. Blood glucose concentration 34mmol/l

c. Prolonged prothrombin time

d. Serum creatine kinase activity 320IU/l

e. Serum aspartate transaminase activity 200IU/l.

25.17 A patient with chronic renal failure is being treated with haemodialysis. On the last four occasions, pre-dialysis serum creatinine concentration has been approximately 450μmol/l and urea approximately 50mmol/l. On the next occasion, 3 days later, pre-dialysis serum creatinine is 448μmol/l, and urea, 68mmol/l. This could be due to

a. Dehydration

b. Development of septicaemia

c. Gastrointestinal haemorrhage

d. Improvement in renal function

e. Increased dietary protein intake.

25.18 A 40-year-old journalist with a history of excessive alcohol ingestion undergoes an 'executive health screen'. Which of the following biochemical results from analysis of serum suggest the presence of an additional problem?

a. Aspartate transaminase activity 60IU/l

b. γ-Glutamyl transpeptidase activity 120IU/l

c. Total cholesterol 9.6mmol/l

d. Triglycerides (fasting) 4.2mmol/l

e. Urate concentration 0.48mmol/l.

25.16 a, b, d.
Paracetamol poisoning alone does not cause carbon dioxide retention nor such severe hyperglycaemia. Respiratory alkalosis and hypokalaemia are more likely if hepatic failure develops. Liver damage causes release of AST, and functional impairment causes prolongation of the prothrombin time. Muscle damage does not occur so that there is no increases in creatine kinase.

25.17 a, b, c, e.
An increase in urea concentration with no change in creatinine can occur in dehydration (due to increased back-diffusion of urea from the tubular fluid), and with increased urea synthesis from dietary protein, blood in the gut or as a result of endogenous protein breakdown. Loss of renal function is irreversible in chronic renal failure and an improvement would anyway be expected to decrease both urea and creatinine.

25.18 c.
Elevations in AST and γ-GT, hyperuricaemia and hypertriglyceridaemia - but not hypercholesterolaemia (unless cirrhosis develops) - can all be due to excessive alcohol intake.

25.19 Analysis of arterial blood from a patient in the
intensive care unit with multiple trauma reveals:
[H^+] 80nmol/l (pH 7.10), P_{CO_2} 7.4kPa, P_{O_2} 8.5kPa,
derived [HCO_3^-] 16mmol/l. The following
statements are correct:

a. The data indicate respiratory compensation for a
non-respiratory alkalosis

b. The patient could have both respiratory and renal
failure

c. The results indicate a failure of renal bicarbonate
reabsorption

d. There is a mixed respiratory and non-respiratory
acidosis

e. There is renal compensation for a primary
respiratory acidosis.

25.20 Thyroid function tests in an elderly woman
admitted to hospital with pneumonia and found to
be in atrial fibrillation reveal: plasma TSH 0.1mU/l,
free T3 2.1pmol/l. The following statements are
correct:

a. Hypothyroidism secondary to pituitary failure cannot
be excluded on the basis of these results

b. Measurement of plasma thyroxine-binding globulin
concentration is indicated

c. She requires treatment with anti-thyroid drugs

d. The most likely diagnosis is the 'sick euthyroid'
syndrome

e. The tests should be repeated when she has
recovered from the acute illness.

25.19 b, d.
There is an acidosis and the elevated $P\text{CO}_2$ indicates that there is a respiratory component; however, $[H^+]$ is higher than would be expected, indicating a non-respiratory component in addition. Respiratory and renal failure could be responsible. There is no reason to suspect impaired renal bicarbonate reabsorption (renal tubular acidosis) specifically, although this cannot be excluded. $[H^+]$ would be lower than expected if there were renal compensation for the respiratory acidosis.

25.20 a, d, e.
The low TSH and free triiodothyronine are typical of the sick euthyroid syndrome though thyroid failure secondary to pituitary diseases cannot be excluded. Thyroid function is difficult to assess in non-thyroidal illness and the tests should be repeated at a later date. Thyroxine treatment is not indicated. Measurement of TBG will not be helpful.

25.21 The following results are found in an adult patient presenting with weight loss, diarrhoea and abdominal discomfort: serum calcium concentration 1.95mmol/l, phosphate 0.6mmol/l, albumin 32g/l, alkaline phosphatase 230iu/l. The following statements are correct:

a. A low plasma concentration of 25-OH vitamin D would be expected

b. An elevated plasma parathyroid hormone would be expected

c. Ionized calcium concentration is likely to be low

d. Malabsorption of fat is likely to be demonstrable

e. The alkaline phosphatase is most likely to be of intestinal origin.

25.22 A newborn infant has ambiguous genitalia; chromosomal studies show the karyotype to be 46XX. The following statements are correct:

a. Congenital adrenal hyperplasia is a possible diagnosis

b. Measurement of plasma 17-OH progesterone is indicated

c. The features are consistent with female pseudo-hermaphroditism

d. The infant is genotypically female

e. Turner's syndrome is a possible diagnosis.

25.23 A 26-year-old man is investigated for infertility associated with a low sperm count. Plasma FSH 18U/l, LH 19U/l, testosterone 4nmol/l. The following statements are correct:

a. A history of mumps may be relevant to his problem

b. Testosterone concentration would be expected to increase in response to administration of clomiphene

c. The data are consistent with primary testicular failure

d. The presence of a pituitary tumour secreting gonadotrophins would explain these data

e. Treatment with testosterone is likely to restore fertility.

25.21 a, b, c, d.
The 'corrected' calcium concentration is low
(2.11mmol/l) suggesting a low ionized calcium and
consistent with vitamin D deficiency. Hypocalcaemia
causes increased secretion of parathyroid hormone
(secondary hyperparathyroidism), and this increases
renal phosphate excretion. Increased alkaline
phosphatase activity in vitamin D deficiency reflects
increased osteoblastic activity.

25.22 a, b, c, d.
The infant is genotypically female; in Turner's
syndrome the usual karyotype is 45X0. Congenital
adrenal hyperplasia can cause virilization of female
infants and plasma 17-OH progesterone concentration
is elevated in the majority of cases.

25.23 a, c.
The low testosterone with elevated gonadotrophins
suggests primary testicular failure; mumps orchitis is a
recognized cause of this. Stimulation of gonadotrophin
secretion with clomiphene has no effect on
testosterone in primary testicular failure.
Gonadotrophin-secreting tumours are excessively rare
and should not cause a low testosterone. Testosterone
does not restore fertility in primary gonadal failure.

25.24 A patient develops a pancreatic fistula following surgery for a pancreatic pseudocyst. Analysis of serum reveals: sodium 134mmol/l, potassium 3.5mmol/l, bicarbonate 14mmol/l, urea 10mmol/l, creatinine 90μmol/l.

    a. A normal anion gap would be expected

    b. He is likely to have a non-respiratory acidosis

    c. Intravenous infusion of sodium chloride would be expected to restore his acid-basis status to normal

    d. Loss of bicarbonate-rich fluid would explain these results

    e. The plasma creatinine concentration indicates a normal glomerular filtration rate.

25.25 A young woman is admitted to the accident and emergency department. She was found semi-conscious at home by her flatmate in the evening, but had been well that morning. On examination she is dehydrated, pyrexial and hyperventilating. Arterial blood [H^+] 50nmol/l (pH 7.30), Pco_2 3.0kPa, bicarbonate 10mmol/l; blood glucose concentration 6.5mmol/l; urine positive to Clinitest, negative for ketones.

    a. Acute alcohol poisoning is a likely diagnosis

    b. Salicylate poisoning would explain these results

    c. The diagnosis is diabetic ketoacidosis

    d. There is evidence of a non-respiratory acidosis

    e. There is evidence of a respiratory alkalosis.

25.26 An elderly woman presents with painless jaundice and weight loss. Serum bilirubin concentration is 282μmol/l, aspartate transaminase 55iu/l, alkaline phosphatase 450iu/l. The following statements are correct:

    a. γ-Glutamyl transpeptidase is likely to be elevated

    b. Concomitant bone disease cannot be excluded

    c. She would be expected to have dark urine

    d. There is a risk of kernicterus

    e. The data are compatible with carcinoma of the head of the pancreas.

25.24 a, b, d.
Non-respiratory acidosis can occur as a result of loss of bicarbonate from a pancreatic fistula. The acidosis is not due to increased production of organic acids and the anion gap should be normal. Plasma creatinine concentration is an insensitive test of renal impairment and can be normal even if the GFR is moderately decreased.

25.25 b, d, e.
She is acidotic; the low $P\text{CO}_2$ precludes this being of respiratory origin and indicates that she also has respiratory alkalosis (which may be compensatory). Alcohol can cause ketoacidosis but she is not ketotic. Blood glucose concentration is normal. Clinitest is not specific for glucose; it tests for reducing substances and salicylate can give a positive result.

25.26 a, b, c, e.
The data suggest an obstructive cause for the jaundice. The excess bilirubin will be conjugated, and not cause kernicterus. Plasma γ-glutamyl transpeptidase activity is frequently elevated in patients with conjugated hyperbilirubinaemia, whatever the cause. Bone disease could be contributing to the high alkaline phosphatase.

25.27 Thyroid function tests are requested on a 60-year-old woman. No clinical details are given on the request form. Serum TSH is 15mU/l, free thyroxine 24pmol/l.

a. A TRH test is required
b. Clinical details are essential for an adequate interpretation of the result
c. Poor compliance with thyroxine treatment is a possible explanation
d. She may have been started on thyroxine treatment only recently
e. The results are typical of the 'sick euthyroid' syndrome.

25.28 A 54-year-old man is referred to a dermatologist with a blistering rash on the face and hands. Porphyrin analysis shows a massively elevated urinary uroporphyrin excretion; no porphobilinogen is detectable in the urine.

a. A history of excessive alcohol consumption would be relevant
b. An excess of red cell porphyrins would be expected
c. Neurological manifestations would be expected
d. The most likely diagnosis is cutaneous hepatic porphyria
e. The urine would be fluorescent.

25.29 A 55-year old man with long-standing chronic obstructive airways disease is admitted to hospital with congestive cardiac failure. The results of arterial blood gas analysis are: $[H^+]$ 46nmol/l (pH 7.34), P_{CO_2} 8.0kPa.

a. A low plasma bicarbonate concentration would be expected
b. Acute retention of carbon dioxide with pre-existing non-respiratory alkalosis cannot be excluded
c. In this context, the most likely explanation for these data is a compensated respiratory acidosis
d. The data cannot be adequately interpreted without the bicarbonate concentration
e. The urine would probably be alkaline.

25.27 b, c, d.
The TSH is elevated, yet free thyroxine is normal. The
TSH result indicates primary thyroid failure. It responds
more slowly than plasma thyroxine to thyroxine
replacement. Treatment may have been started
recently or the patient may be non-compliant. It can be
concluded from the elevated TSH that the response to
TRH would be exaggerated—the test will not provide
useful additional information.

25.28 a, d, e.
Cutaneous hepatic porphyria is a photosensitizing
porphyria and is associated with high urinary
uroporphyrin excretion, causing the urine to be
fluorescent. Porphyrin precursors are not produced in
excess; neurological manifestations are not a feature
of this porphyria. It can be acquired, and is particularly
associated with a high alcohol intake. Red cell
porphyrins are normal in this condition.

25.29 b, c.
He is slightly acidotic, with a high P_{CO_2}, indicating a
respiratory acidosis. The $[H^+]$ is not as high as would
be predicted were this an acute disturbance, indicating
that a metabolic alkalosis (most probably
compensatory in view of the history) is present.
Plasma bicarbonate concentration would be expected
to be high in either case, while increased hydrogen ion
excretion would be expected to cause an acid urine.

25.30 Three days after an abdominal operation,
biochemical analysis of a patients serum reveals:
urea 9.6mmol/l, creatinine 90μmol/l, calcium
2.72mmol/l, phosphate 1.25mmol/l, albumin 51g/l.
No biochemical abnormality had been present
preoperatively. These data suggest
a. Acute phase response
b. Acute tubular necrosis
c. Dehydration
d. Normal ionized calcium concentration
e. Primary hyperparathyroidism.

25.31 A 38-year-old man is screened for
hypercholesterolaemia after his elder brother
suffers a myocardial infarct. Serum cholesterol
concentration (non-fasting) is 13.0mmol/l,
triglycerides 1.9mmol/l.
a. Analysis should be performed on a fasting sample
before any action is taken
b. His thyroid function should be checked
c. Lipid-lowering treatment is unlikely to be required if
there are no other risk factors for coronary artery
disease
d. Serum electrophoresis would be likely to show an
increase in the β-lipoprotein band.
e. The most likely diagnosis is familial
hypercholesterolaemia.

25.30 c, d.
'Corrected' calcium is normal; plasma albumin concentration decreases during the acute phase response but can be increased by dehydration. The slightly elevated urea and normal creatinine are compatible with dehydration: much higher concentrations would be expected in acute tubular necrosis.

25.31 b, d, e.
A cholesterol this high in a young man with a family history of ischaemic heart disease is most likely to be due to familial hypercholesterolaemia. Hypothyroidism can also cause hypercholesterolaemia and, however unlikely, should always be excluded. Drug treatment is usually required to bring the cholesterol to within acceptable limits. Recent food intake does not affect plasma cholesterol concentration. The excess cholesterol is in LDL, which has β-mobility on electrophoresis.

25.32 A patient on parenteral feeding is having continuous 24-hourly collections of urine made to assess nitrogen excretion. On a constant input of 14g daily, values for urea excretion on successive days are: 400mmol, 480mmol, 390mmol, 50mmol. Serum urea is unchanged, and urine volume is appropriate for fluid input.

   a. Acute renal failure is the probable cause of the decreased urea excretion.

   b. Bacterial contamination of the urine could explain the result on the fourth day

   c. Laboratory error could explain the result on the fourth day

   d. The approximate nitrogen excretion (as urea) on the first three days is 7.5g

   e. The data on the first three days suggest that she is in positive nitrogen balance.

25.33 A baby boy, born at term by normal vaginal delivery and initially well, develops tachypnoea on the third day of life. Arterial blood [H$^+$] 50nmol/l (pH 7.30), Pco$_2$ 3.3kPa. Urine is negative for reducing substances. He becomes progressively more acidotic and next day plasma ammonia concentration is measured and found to be very high. Possible diagnoses include

   a. Congenital hypothyroidism

   b. Galactosaemia

   c. Organic acidaemia

   d. Respiratory distress syndrome

   e. Urea cycle defect.

25.32 b, c, e.
The unchanged serum urea effectively excludes renal
failure. One mole of urea $\{CO(NH_2)_2\}$ contains one
mole of nitrogen (N_2, molecular weight 28).
Approximate nitrogen excretion is at least equal to the
nitrogen excreted as urea ($390/1000 \times 28 = 10.9$g).
Allowing 2g for non-urea losses, the data suggest that
she is in positive nitrogen balance (input 14g, output
13g). Bacteria can metabolise urea to ammonia. When
a result is unexpected, a laboratory error should be
considered, but never assumed.

25.33 c, e.
He has a partially compensated non-respiratory
acidosis; in conjunction with hyperammonaemia this is
compatible with a urea cycle enzyme defect or an
organic acidaemia. Respiratory distress syndrome
would cause a respiratory or mixed acidosis. The liver
disease associated with galactosaemia can cause
hyperammonaemia but a positive urine test for
reducing substances would be expected. Congenital
hypothyroidism does not present in this way.

25.34 A 30-year-old man with known chronic liver disease is admitted to hospital following a massive haematemesis. This is controlled with a Sengstaken tube. During the next 2 days, considerable quantities of blood and fluid are aspirated from the stomach, and he is given blood transfusions and intravenous Ringer's lactate solution. He becomes encephalopathic. Arterial blood: [H$^+$] 20nmol/l (pH 7.71), Pco$_2$ 3.8kPa:

a. There is evidence of a respiratory alkalosis

b. The low Pco$_2$ is a compensatory phenomenon

c. A high plasma bicarbonate concentration would be expected

d. There is a mixed respiratory and non-respiratory alkalosis

e. The [H$^+$] is incompatible with life and the results must be erroneous.

25.35 An elderly woman is brought to hospital by ambulance, having been found at home by a neighbour in an unrousable state. On examination, she is very dehydrated. Respiration is normal; her urine is positive for glucose, negative for ketones. Biochemical analysis shows: plasma sodium concentration 150mmol/l, potassium 4.8mmol/l, bicarbonate 20mmol/l, urea 45mmol/l, creatinine 180μmol/l, blood glucose 62mmol/l:

a. A plasma osmolality of approximately 400mmol/kg would be expected

b. She is severely acidotic

c. The most likely diagnosis is hyperosmolar, non-ketotic hyperglycaemia

d. The results for urea and creatinine suggest that she has been consuming a high protein diet

e. The sodium concentration suggests a high habitual salt intake.

25.34 a, c, d.
He is alkalotic with a low P_{CO_2}, indicating a respiratory alkalosis, but the $[H^+]$ is lower than predicted indicating a metabolic component also.
Hyperventilation lowers the P_{CO_2} as a compensatory change in non-respiratory acidosis, not alkalosis.
Bicarbonate is increased in non-respiratory alkalosis.
$[H^+]$ is unusually low, but not incompatible with life.

25.35 a, c.
The calculated osmolality is as given; bicarbonate is only slightly low, suggesting only a mild disturbance of acid-base homoeostasis. These results are typical of hyperosmolar, non-ketotic hyperglycaemia; excessive water loss causes hypernatraemia and dehydration causes renal impairment, the urea often being much higher than creatinine because of back-diffusion from the renal tubules.

25.36 The following measurements are made for the calculation of an elderly female diabetic patient's creatinine clearance: 24h urine volume, 1.44l; plasma creatinine concentration 100μmol/l; urine creatinine concentration 6.6mmol/l.

a. Clinical features of renal impairment would be expected

b. Plasma creatinine alone indicates impaired renal function

c. Plasma potassium concentration should be measured urgently

d. The data suggest renal impairment

e. There is reason to suspect an incomplete urine collection.

25.37 The value of a new diagnostic test is assessed in a group of patients known to have the disease and in an equal number of healthy controls. The test is performed once in each individual. Out of a total of 200 tests 106 positive results are recorded; 85 of the individuals known to have the disease test positive. The following conclusions are correct:

a. There are 15 false negative results

b. The specificity of the test is 85%

c. The sensitivity of the test is 94%

d. The efficiency of the test is 82%

e. In this study, the predictive value of a positive result is approximately 84%

25.36 d.
Creatinine clearance is $(6600 \times 1440/1440)/100 =$ 66ml/min. This suggests renal impairment but the plasma creatinine is normal. Clinical evidence of renal impairment is unlikely to be present with a glomerular filtration rate this high and significant hyperkalaemia is unlikely. The urine volume is within normal limits - the patient is an elderly woman who might be expected to have a small muscle bulk and thus endogenous creatinine production and excretion.

25.37 a, d.
Analysis of the data shows: 85 true positives, 21 false positives, 79 true negatives, 15 false negatives. Thus:
sensitivity = 85%
specificity = 79%
efficiency = (85+79)/100% = 82%
predictive value
for positive test = 85/(85+21)% = 80.2%

25.38 An insulin-dependent diabetic patient is admitted to hospital after collapsing during a 50-mile cycle race, which he had entered to raise money for charity. His blood glucose concentration is 0.5mmol/l. He is given 50% dextrose intravenously and makes a rapid recovery. Plasma creatine kinase activity is normal on admission but 12 hours later is reported as 240IU/l. The following statements are correct:

   a. A very low glycated haemoglobin (HbA$_{1c}$) level would be expected

   b. Measurement of creatine kinase MB isoenzyme activity on the sample at 12 hours would be diagnostically useful

   c. Measurement of hydroxybutyrate dehydrogenase activity on the sample at 12 hours would be diagnostically useful.

   d. The creatine kinase result could be a direct result of exercise

   e. The creatine kinase result indicates that his collapse was due to myocardial infarction.

25.39 A male infant is born at term to parents who are first cousins. He refuses feeds and 12 hours after birth is noticed to be 'jittery'; blood glucose concentration measured by reagent stick is <1.0mmol/l. Before any action can be taken, he develops focal seizures. Investigations reveal: arterial blood [H$^+$] 50 (pH 7.30), Pco$_2$ 3.4kPa; urine negative for reducing substances, positive for ketones.

   a. Glycogen storage disease type I is a possible diagnosis

   b. Hereditary fructose intolerance is a likely diagnosis

   c. Measurement of plasma lactate concentration would be diagnostically useful

   d. The acidosis is respiratory in origin

   e. The presence of an elevated plasma urate concentration would suggest a specific diagnosis.

25.38 b, d.
Glycated haemoglobin is stable once formed and is not decreased by a single episode of hypoglycaemia. Severe exercise can increase creatine kinase activity to this extent; myocardial infarction is a possibility, but not a certainty. Myocardial infarction would be suggested if >5% of the creatine kinase was the MB isoenyzme; hydroxybutyrate dehydrogenase activity does not increase in the first 12 hours following an infarct.

25.39 a, c, e.
He is hypoglycaemic and ketotic with a partially compensated non-respiratory acidosis; this is typical of glycogen storage disease type I (the acidosis is a lactic acidosis), in which hyperuricaemia is often present. Hereditary fructose intolerance causes hypoglycaemia on exposure to fructose (and sucrose) and is associated with fructosuria, causing a positive test for reducing substances.

25.40 **A 4-year-old girl is investigated for short stature. Investigations reveal: plasma calcium concentration 1.70mmol/l, phosphate 3.8mmol/l, albumin 39mmol/l, alkaline phosphatase 245IU/l, creatinine 52μmol/l. The following statements are true:**

a. Clinical evidence of rickets would be expected

b. Growth failure is probably secondary to renal failure

c. Hypoparathyroidism is a possible diagnosis

d. Measurement of urinary cyclic AMP excretion following intravenous infusion of parathyroid hormone could be diagnostically useful

e. The data suggest vitamin D deficiency.

25.40 c, d.
The creatinine concentration does not suggest renal failure. Vitamin D deficiency causes hypocalcaemia and hypophosphataemia (due to secondary hyperparathyroidism); hyperphosphataemia suggests hypoparathyroidism. Also, alkaline phosphatase is normal for a child of this age. Measurement of urinary cyclic AMP excretion following PTH infusion can distinguish between true and pseudohypoparathyroidism, which affect plasma calcium and phosphate concentrations similarly.

Adult reference ranges	
acid phosphatase: total prostatic	4–11IU/l <4IU/l
adrenocorticotrophic hormone (ACTH): at 0900h	10–80ng/l
albumin	35–50g/l
aldosterone: recumbent	100–500pmol/l
alkaline phosphatase	30–90IU/l
alphafetoprotein (AFP)	<10kU/l
ammonia	10–47μmol/l
amylase	<300IU/l
aspartate transaminase (AST)	10–50IU/l
bicarbonate (total CO_2)	22–30mmol/l
bilirubin: total	3–20μmol/l
calcium	2.2–2.6mmol/l
carbon dioxide (P_{CO_2}): arterial blood	4.5–6.0kPa (35–46mmHg)
cholesterol: total high density lipoprotein (HDL) low density lipoprotein (LDL)	<5.2mmol/l* >1.2mmol/l* <3.5mmol/l*
* indicates ideal values.	
copper	12–19μmol/l
cortisol: at 0900h at 2400h	140–690nmol/l <100nmol/l

Adult reference ranges	
creatine kinase (total)	<90IU/l
creatinine	60–110μmol/l

follicle-stimulating hormone (FSH):
  adult males
  females:
    follicular phase
    post-menopausal
	
	2–10 U/l
	
	2–8 U/l
	>15U/l

glucose: fasting	2.8–6.0mmol/l
γ-glutamyl transpeptidase (γGT)	<60IU/l

growth hormone:
  following glucose load
  following stress
	
	<2mU/l
	>20mU/l

haemoglobin: males females	13–18g/dl 12–16g/dl
hydrogen ion: arterial blood	35–46nmol/l (pH 7.36–7.44)
hydroxybutyrate dehydrogenase (HBD)	<250IU/l
insulin: fasting in hypoglycaemia	3–15mU/l <3mU/l

luteinizing hormone (LH):
  adult males
  adult females:
    follicular phase
    post-menopausal
	
	2.0–10U/l
	
	2.0–10U/l
	>20U/l

magnesium	0.7–1.0mmol/l

Adult reference ranges	
osmolality	280–295mmol/kg
oxygen (Po_2): arterial blood	11–15kPa (85–105mmHg)
parathyroid hormone (N-terminus)	10–65pg/ml
phosphate	0.8–1.4mmol/l
potassium	3.6–5.0mmol/l
prolactin	50–350mU/l
protein: total	60–80g/l
renin (plasma renin activity, PRA): recumbent	1.2–2.4pmol/h/ml
sodium	135–145mmol/l
testosterone: adult males	9–30nmol/l
adult females	0.5–2.5nmol/l
thyroid-stimulating hormone (TSH, thyrotrophin)	0.3–4.0mU/l
thyroxine (T4): total	60–150nmol/l
free	9–26pmol/l
triglyceride: fasting	0.4–1.8mmol/l
triiodthyronine (T3): total	1.2–2.9nmol/l
free	3.0–8.8pmol/l
urea	3.3–6.7mmol/l
uric acid	0.1–0.4mmol/l
zinc	12–20μmol/l

Answers

1.1 b, c, d.	**2.14** a, e.	**4.13** b, d.
1.2 b.	**2.15** a, b, c, e.	**4.14** c.
1.3 b, c.	**2.16** a, b, c, d, e.	**4.15** a, b, c, d.
1.4 a, d, e.	**2.17** a, b, c, e.	**4.16** b, e.
1.5 b, e.	**3.1** c, d, e.	**4.17** a, b.
1.6 b, d.	**3.2** a, b, c.	**4.18** a, b, d, e.
1.7 a, b, e.	**3.3** b, e.	**4.19** c, d, e.
1.8 a.	**3.4** b, c.	**5.1** a, b, e.
1.9 a.	**3.5** a, c, d.	**5.2** a, c, d, e.
1.10 a, b, c, d, e.	**3.6** a, b, c, d.	**5.3** b.
2.1 e.	**3.7** b, d, e.	**5.4** a
2.2 b, c.	**4.1** a, c, d, e.	**5.5** b, d.
2.3 a, b, d, e.	**4.2** b, d.	**5.6** a, b, c.
2.4 c, e.	**4.3** d.	**5.7** a, b, d.
2.5 a, b, e.	**4.4** b, c, e.	**5.8** b, e.
2.6 c.	**4.5** a, c.	**5.9** a, c, d, e.
2.7 c, e.	**4.6** b, d, e.	**5.10** a, d.
2.8 b.	**4.7** b, e.	**5.11** b, d.
2.9 b, c, e.	**4.8** a, c, e.	**5.12** e.
2.10 a, b, c, e.	**4.9** a, b, c, d.	**5.13** c.
2.1 d.	**4.10** d, e.	**5.14** c, d, e.
2.12 c, d, e.	**4.11** b, c, d.	**5.15** a, b, c, d, e.
2.13 a, d.	**4.12** a, c, e	**5.16** a, c.

5.17 a, d, e.	**7.6** b, d, e.	**9.2** a, b, c.
5.18 b, d.	**7.7** a, b.	**9.3** b, c, e.
5.19 a, d, e.	**7.8** e.	**9.4** a, e.
5.20 b, c, d, e.	**7.9** c, d.	**9.5** a, b, d.
6.1 a, b, c, e.	**7.10** a, b, e.	**9.6** b, d, e.
6.2 a, b, e.	**7.11** b, d, e.	**9.7** b, d, e.
6.3 b, e.	**7.12** a, b, e.	**9.8** d.
6.4 b, e.	**7.13** a, b, d.	**9.9** a, b, e.
6.5 d.	**7.14** d, e.	**9.10** a, c, e.
6.6 a, e.	**7.15** a, e.	**9.11** a, b, c, d.
6.7 c, e.	**7.16** c, e.	**9.12** a, c, e.
6.8 a.	**7.17** a, b, d, e.	**9.13** e.
6.9 b, c.	**8.1** c, d, e.	**9.14** a, c, d, e.
6.10 b, c, d.	**8.2** d.	**9.15** b, d.
6.11 a, b, e.	**8.3** c, d, e.	**9.16** a, b, e.
6.12 a, b, c, d, e.	**8.4** a, b, c.	**9.17** a, b, e.
6.13 c, d, e.	**8.5** b, e.	**10.1** a, d.
6.14 c, d, e.	**8.6** a, c, d,e.	**10.2** b, c, d, e.
6.15 a, b, d, e.	**8.7** c, e.	**10.3** a, b, c.
6.16 c, d, e.	**8.8** b, e.	**10.4** c, e.
6.17 a.	**8.9** a, b, d.	**10.5** a, b, c, d.
6.18 a, b.	**8.10** c.	**10.6** a, c.
6.19 c, d.	**8.11** a, b, c, d.	**10.7** e.
7.1 c, d, e.	**8.12** b, e.	**10.8** c, d.
7.2 a, b, c, e.	**8.13** a, d.	**10.9** a, b, c, d, e.
7.3 a, c, d.	**8.14** b, d.	**10.10** a, d.
7.4 b, e.	**8.15** a, b, c, d, e.	**10.11** b, d, e.
7.5 a, b, c, e.	**9.1** a, b.	**10.12** a, d.

10.13 b, c, d, e.	**12.10** a, e.	**14.8** a, b, d, e.
10.14 a, b, d.	**12.11** a, b, d, e.	**14.9** b, d, e.
10.15 a, e.	**12.12** a, c, d.	**14.10** a, c.
10.16 b, d, e.	**12.13** a, b, c, d.	**14.11** a, b.
10.17 a, b, e.	**12.14** e.	**14.12** c, e.
11.1 a, b, c.	**12.15** a, c, d, e.	**14.13** a, c, e.
11.2 b, c, d, e.	**12.16** a, b, c, e.	**14.14** a, b, c, d, e.
11.3 a, c, d, e.	**12.17** a, b, c, d.	**14.15** a, b, d, e.
11.4 c.	**13.1** a, b.	**14.16** a, c, d.
11.5 a, c, d.	**13.2** a, c, d.	**14.17** a.
11.6 a, b, d, e.	**13.3** a, b, e.	**15.1** a, b, c, d, e.
11.7 a, b, c, d, e.	**13.4** b, d, e.	**15.2** e.
11.8 c.	**13.5** a, b, c, e.	**15.3** c, e.
11.9 b, c.	**13.6** b, c, e.	**15.4** a, b, d, e.
11.10 a, b, c, d.	**13.7** d.	**15.5** b, d, e.
11.11 b, c, d.	**13.8** a, c, e.	**15.6** a, c, e.
11.12 c, e.	**13.9** d, e.	**15.7** a, b, d.
11.13 b, c.	**13.10** b, d, e.	**15.8** d, e.
11.14 c, d.	**13.11** a, c, d, e.	**15.9** b, d.
12.1 a, c, d.	**13.12** d.	**15.10** a, b, c.
12.2 a, b, d.	**13.13** a, d.	**15.11** a, b, c.
12.3 a, b, d.	**14.1** c, e.	**15.12** c, d.
12.4 a.	**14.2** a, d.	**15.13** e.
12.5 a, b, c,e.	**14.3** a, b, d.	**15.14** b, e.
12.6 a, b, e.	**14.4** b, c, d, e.	**15.15** a, b, d.
12.7 a, b, e.	**14.5** b, c.	**15.16** a, b, c, e.
12.8 a.	**14.6** b, d, e.	**16.1** b, d, e.
12.9 e.	**14.7** a, c, e.	**16.2** a, d.

16.3 c, d, e.	**17.13** b, c, d, e.	**19.11** b, d.
16.4 a, b, c, d, e.	**17.14** c, e.	**19.12** b, e.
16.5 c, e.	**17.15** d, e.	**19.13** a, c, d.
16.6 c, d, e.	**18.1** a, b, c, d, e.	**20.1** b, c, e.
16.7 a, b, e.	**18.2** b, e.	**20.2** a, b.
16.8 a, b, c, e.	**18.3** c, d, e.	**20.3** b, c.
16.9 a, b, c, e.	**18.4** b, d.	**20.4** a, b, c, d.
16.10 b, c, e.	**18.5** a, b, d, e.	**20.5** c, e.
16.11 b, e.	**18.6** a, b, d.	**20.6** a, b, d.
16.12 c, e.	**18.7** b, c.	**20.7** a, b, c, e.
16.13 b, c, d.	**18.8** b, c, d.	**20.8** a, b, c.
16.14 a, e.	**18.9** b, d.	**21.1** a, b, c, d, e.
16.15 a, b, c, d.	**18.10** a, b, e.	**21.2** b, c, d, e.
16.16 b, c, d.	**18.11** c.	**21.3** e.
16.17 c.	**18.12** b, c, d.	**21.4** a, e.
16.18 c, e.	**18.13** c, e.	**21.5** e.
17.1 a, b, c, d, e.	**18.14** a, b.	**21.6** a, b, d.
17.2 d.	**18.15** a, b, c.	**21.7** a, d, e.
17.3 c, e.	**19.1** a, b, d, e.	**21.8** e.
17.4 a, b, e.	**19.2** a, b, c, d, e.	**21.9** a, b, c, d.
17.5 b, c, d.	**19.3** a, c, d.	**21.10** c, d, e.
17.6 a, d.	**19.4** a, b, e.	**21.11** a, b, c.
17.7 a, e.	**19.5** d.	**21.12** a, b, c, e.
17.8 a, c.	**19.6** a, c, d.	**21.13** a, b, d.
17.9 a, b, e.	**19.7** a, b, e.	**21.14** a, c, d, e.
17.10 a, c, d, e.	**19.8** b, d.	**21.15** a, b, d.
17.11 a, b, c, d.	**19.9** b.	**21.16** a, b, c, d.
17.12 a, c, d.	**19.10** b, d, e.	**21.17** a.

22.1 a, b.	**23.1** a, b, e.	**24.1** a, b.
22.2 a, b, c.	**23.2** b, c, e.	**24.2** a, b, d, e.
22.3 a, e.	**23.3** a, b, c, d, e.	**24.3** a, b, c, d.
22.4 a, b, c.	**23.4** a, b.	**24.4** a, c, e.
22.5 a, d, e.	**23.5** b, d.	**24.5** b, d.
22.6 a, b.	**23.6** b, c, d.	**24.6** a, b.
22.7 b, d, e.	**23.7** a, e.	**24.7** a, c, d, e.
22.8 a, c.	**23.8** b, c, d.	**24.8** b, c, d, e.
22.9 a.	**23.9** b, e.	**24.9** d.
22.10 a, b.	**23.10** a, b, c, d, e.	**24.10** a, b, c, d, e.
22.11 b, c, d, e.	**23.11** e.	**24.11** b, c.
22.12 a, b, c, d, e.		
22.13 a, c.		
22.14 a, b, d, e.		
22.15 b, d, e.		
22.16 c, d, e.		
22.17 a, b, c, e.		